Adonijah Strong Welch

The Teachers' Psychology

Adonijah Strong Welch

The Teachers' Psychology

ISBN/EAN: 9783337366308

Printed in Europe, USA, Canada, Australia, Japan

Cover: Foto ©berggeist007 / pixelio.de

More available books at **www.hansebooks.com**

The · Teachers' · Psychology

A · TREATISE · ON · THE · INTEL-
LECTUAL · FACULTIES, · THE
ORDER · OF · THEIR · GROWTH,
AND · THE · CORRESPONDING
SERIES · OF · STUDIES · BY
WHICH · THEY · ARE · EDU-
CATED.

BY

A. S. WELCH, LL.D.,

Professor of Psychology, Iowa College of Agriculture and Mechanic
Arts, Ames, Iowa.

New York and Chicago:
E. L. KELLOGG & CO.
1893.

"*TALKS ON PSYCHOLOGY,*" *by Dr. A. S. Welch, consists of an <u>outline</u> of the Principles of Psychology as related to teaching, designed for study in schools and teachers' institutes. Cloth, 16mo, 136 pp., 50 cents. E. L. Kellogg & Co., New York and Chicago.*

Copyright, 1889,
E. L. KELLOGG & CO.

PREFACE.

THE TEACHER'S PSYCHOLOGY has been written for the purpose of meeting an urgent need now widely felt by the teachers themselves. A complete mastery of the branches taught is no longer regarded as an adequate preparation for the teacher's work. For true teaching not only guides the efforts of the pupil in acquiring *knowledge,* but incites the *kind of efforts* that contribute to his intellectual strength. This double purpose of genuine instruction demands a corresponding increase in the range of the teacher's attainments. His knowledge must embrace not simply the facts he teaches, but also the faculties he trains. He must know the mind with which he deals: its impulses; its spontaneities; its dominant activities, and the invariable order in which its powers are developed.

Now, this knowledge of mind and its laws may be gained by a persistent self-scrutiny, which can be rightly guided and materially helped by a book judiciously prepared for the purpose. And such a book I have endeavored to write after studying and teaching, for some thirty years, the subjects of which it treats.

Let me earnestly urge that you who read the TEACHER'S PSYCHOLOGY, apply its matter simply as an incentive to the study of your own mental operations.

Test every manifestation of mind described by producing it yourself. Discriminate and realize, in your own efforts, the object, action, and product of each faculty under consideration. Faithfully repeat these experiments in self-scrutiny until you attain thereby a knowledge of psychology which will serve as an infallible guide in the duties of your chosen profession.

A. S. WELCH.

AMES, IOWA.

CONTENTS.

CHAP. I.—Introduction—Terms Defined and their Meanings Illustrated.

SEC.
1. The Value of definite Terms, 1.
2. Mind, 2.
3. Knowing, 2.
4. Feeling, 2.
5. Willing, 3.
6. The Will—Effort-making, 3.
7. Spontaneity—Self-movement, 3.
8. What is a Faculty? 4.
9. The Object, Action, and Product of a Faculty, 4.
10. Attention—The Centring of Effort, 5.
11. What is a Sensation? 5.
12. Sense-perception—The Gathering of Concepts, 5.

SEC.
13. A Product, 6.
14. Memory—Storing Concepts, 6.
15. Conception—Holding Concepts, 7.
16. Analysis—Dividing Concepts, 8.
17. Abstraction—Transforming Concepts, 8.
18. Imagination—Building Concepts, 9.
19. Classification—Grouping Concepts, 10.
20. Judgment—Connecting Concepts, 11.
21. Reasoning—Deriving Concepts, 11.

CHAP. II.—Mind and its Three Manifestations.

SEC.
22. Only two Forms of Existence, 14.
23. The Manifestations of the Mind of three kinds, 15.
24. Knowledge, 15.
25. The Feelings, 15.
26. Knowledge precedes Feeling, 16.
27. Sensations, 16.
28. Sensation either Pleasant or Painful, 17.
29. The Appetites, 17.
30. Selfish Feelings—Egoism, 18.
31. Selfish Feelings that are Pleasant, 19.
32. The Social Feelings—Altruism, 19.
33. The Emotions, 20.
34. The Emotion of Beauty, 21.
35. Emotions of Sublimity, 21.
36. Love of Knowledge, 22.
37. The Moral Sense—Conscience, 22.
38. The Religious Emotion, 23.
39. Every Feeling attended by Pleasure or Pain, 23.

SEC.
40. The Gratification of any Feeling uniformly Pleasant, 24.
41. The Desires, 24.
42. The Will, 24.
43. Will no immediate Control over the Feelings, 25.
44. Will uniformly preceded by Desire, 25.
45. Motive—Freedom of Choice, 26.
46. The True Motive, 27.
47. High Moral Character, 28.
48. The Order of Sequence, 28.
49. The Order of our Mental Operations, 28.
50. Rapidity of our Mental Operations, 29.
51. Names which designate the Triple Phenomena, 29.
52. Consciousness, 29.
53. Consciousness a present Knowledge, 30.
54. Consciousness recognizes the Mind as the Originator of its own Acts, 31.
55. Consciousness identical with each Mental Act, 31.

CHAP. III.—On the Intellect—The Senses.

SEC.
56. The Purpose of the Senses, 35.
57. The Organs of the Senses, 35.
58. Sense Perception and Sensation distinguished, 36.
59. Contrast of Sensation and Perception in Consciousness, 37.
60. Sensations and Perceptions as compared in Memory, 39.
61. Names of Sensations and Perceptions compared, 40.
62. The Purposes which the Senses Subserve, 42.
63. Relation of Sense Perception to the Mind, 42.
64. The Pleasures of Sensations compared with those of Sense Perception, 43.

SEC.
65. Pleasures derived from Sensation, 43.
66. Pleasures derived from Sense Perception, 45.
67. The Six Senses—Order of their Growth, 47.
68. Sense Perception later in Activity, 47.
69. The order of growth summed up, 48.
70. The Intellectual Senses—Their Order of Development, 48.
71. The Hand Teaches the Eye, 48.
72. The Ear and Eye Compared, 50.
73. Specimens of Experience in the three kinds of Sense Perception, 57.

CHAP. IV.—Internal Perception.

SEC.
74. Internal Perception Defined, 56.
75. Attention, Observation, Reflection, 58.

SEC.
76. Observation, 59.
77. Reflection, 59.
78. Order of its Development, 60.

CHAP. V.—Memory.

SEC.
79. Order of its Activity, 62.
80. Varieties of Memory, 62.
81. Value of Memory, 63.
82. Memory includes three Acts, 64.
83. Order of Action, 64.
84. Attention necessary to Acquisition, 66.
85. Selection in Acquiring, 66.
86. Special Inclination, 67.
87. Beauty Helps Acquisition, 68.
88. Novelty Helps Acquisition, 69.
89. Strong Feelings Help Acquisition, 69.
90. Retention—Possible Knowledge, 70.
91. Possible Knowledge naturally Transient, 70.
92. Possible Knowledge—How Preserved, 71.
93. Two Methods of Renewal, 72.
94. Values of Reviews, 73.
95. The Retentive Capacity the Storehouse of all Knowledge, 73.

SEC.
96. Recollection the Act of Recalling, 73.
97. The Laws of Association, 74.
98. Associations of Time, 74.
99. The Association of Place, 75.
100. Time and Place in Conjunction, 77.
101. Association of Whole and Parts, 77.
102. Sign and Thing Signified, 78.
103. Cause and Effect, 79.
104. Resemblance, 80.
105. Contrast, 81.
106. Every Intellectual Act a Spontaneity in the Beginning, 81.
107. All Unconscious Modifications of the Mind Spontaneous, 82.
108. How the Will Controls the Thoughts in Voluntary Recollection, 83.
109. Examples of Controlling Thought by the Will, 84.
110. Contrast in Kinds of Thinking, 85.

CHAP. VI.—Conception.

SEC.
111. Conception, defined and illustrated, 90.
112. Concepts may be Modified by Imagination, 91. [91.
113. Concepts of Sight, Reverie, etc.,
114. Sight Concepts superior in Number and Distinctness, 92.
115. Sight Concepts Assumed as Standards, 92.

SEC.
116. Concepts of Touch and Hearing, 93.
117. The Effect of Feeling on Conception, 94.
118. Ideal Presence, 94.
119. The Teacher's Concepts should be Distinct, 95.
120. What Contitutes the Power of Description, 96.

CHAP. VII.—Analysis.

SEC.
121. Analysis Defined and Illustrated, 98.
122. Difference between Scientific and ordinary Analysis, 99.

SEC.
123. The Concept depends on the Precept, 100.
124. Analysis of an Object under Scrutiny of the Senses, 101.

CHAP. VIII.—Abstraction.

SEC.
125. The Nature of the Process, 104.
126. Early Abstraction, 105.

SEC.
127. All Abstract Ideas have a like Origin, 105.
128. Use of Abstract Concepts, 106.

CHAP. IX.—Imagination.

SEC.
129. Concrete Synthesis, 108.
130. Imagination not the Origin of its own Materials, 109.
131. Distinction between the Sense Concept and the Image Concept, 109.
132. Points of Resemblance between Image and Sense Concept, 110.
133. Imagination often distorts Facts, 110.
134. Effect of Passion on Imagination, 110.

SEC.
135. The Eye furnishes the principal Material for Imagination, 110.
136. Wide Range of Imagination—Expression, 111.
137. Influence of Imagination on Character, 112.
138. Imagination in the Arts, 112.
139. Imagination in the Useful Arts, 112.
140. Imagination in the Fine Arts, [113.
141. Genius, 115.
142. Taste, 115.
143. Language, 116.

CHAP. X.—Classification.

SEC.
144. The Mental Acts which precede Classification, 119.
145. The Mental process of Classification, 121.
146. Desultory Classification, 122.
147. Scientific Classification, 123.
148. Perfect Classification, 124.

SEC.
149. Classification based on Form, 150. Definition, 128. [126.
151. Scientific Classification of Concrete Objects, 129. [130.
152. Comprehension and Extension,
153. Office of Language in Classification, 130.

CHAP. XI.—Judgment.

SEC.
154. The Uniform Succession which precedes Class Judgment, 134.
155. Comparison of the fundamental Act in Thinking, 135.

SEC.
156. Class Judgment in Especial, 136.
157. The Proposition, 139.
158. Judgment in Extension and Comprehension, 142.

CHAP. XII.—Reasoning.

SEC.
159. Reasoning, 145.
160. All Reasoning Deductive or Inductive, 145.
161. Reasoning a Comparison of Three Concepts, 146.
162. The Syllogism, 148.
163. Two Kinds of Deductive Reasoning, 148.
164. Deductive Reasoning in Comprehension, 149.

SEC.
165. The General Truths from which deductive Reasoning proceeds, 150. [same Level, 151.
166. Premise and Conclusion on the
167. Self-evident Truths often used as the Basis of Deductive Reasoning, 151. [suppressed, 152.
168. Premises in the Reasoning often
169. Inductive Reasoning, 152.
170. Hasty Induction, 153.

CHAP. XIII.—The Invariable Series of Mental Acts that in the Growth of Mind Begin with the Senses and End in Reasoning.

SEC.
171. The Tri-unity of Power, Act, and Product. 155.
172. Distinction between the Object of a Faculty and its Product, 155.
173. Tabular view of the Succession of Powers, Objects, Acts, and Products, 156. [157.
174. In what our Knowledge consists,

SEC.
175. Maturer Faculties not confined to the Order of Growth, 158.
176. The Successive Intellectual Acts when the class is a familiar one, 158.
177. The Succession of Intellectual Acts when exerted on Objects that are unfamiliar, 161.

CHAP. XIV.—Intuition.

SEC.
178. Intuitive Ideas, 163.
179. Concepts not gained through external or internal Perception, 164.
180. The Concept of Space, 165.
181. The Concept of Time, 167.
182. The Concept of Cause, 168.
183. Substance, 170.
184. The Conscious Operations of Mind Necessitate the Conceptions of Time, Cause, and Substance, 171.
185. Time Suggested by the Manifestations of the Mind, 171.
186. Cause likewise associated with all the Operations of the Mind, 172.

SEC.
187. Mental Operations suggest the Concept of Substance, 172.
188. Intuition, 172.
189. Intuition—Its Early Activity, 173.
190. Terms which Designate the Objects of the Originating Faculties, 174.
191. Concepts of Phenomena, the Chronological Antecedent—Those of Noumena—The Chronological Consequent, 175.
192. Further Facts respecting our necessary Ideas, 178.
193. Our Knowledge of Necessary Truths, 179.

CHAP. XV.—Education—What Is It and How Attained.

SEC.
194. We have learned, 183.
195. Education—What is it? 184.
196. Physical Education, 184.
197. Moral Education, 184.
198. Intellectual Education, 184.
199. The Hand, 185.
200. The Faculty of Conception, 185.
201. Imagination the Concept-builder 186.
202. The Classifying Faculty, 186.
203. Judgment and Reasoning, 187.
204. Every Faculty or Group of Faculties must be trained to supply complete Materials for the Action of the Faculty that follows it in the Series, 187.
205. The Environment Supplies Primarily the Objects by which the Faculties are incited to Action, 189.

SEC.
206. Selection of Objects that elicit Effort, 189.
207. Studies may be selected and arranged in a Series that shall accord with the Series of Unfolding Faculties, 189.
208. Every Intellectual Faculty can be disciplined only by strenuous and reiterated Efforts of Attention directed to its Object, 191.
209. Reiteration of strenuous Efforts begets finally the Habit of Attention which results in Culture, 192.
210. Character of Concepts gathered by disciplinary Efforts, 193.

CHAP. XVI.—The Special Means of Training Each Faculty in the Order of its Growth.

SEC.
211. Principles in Education, 195.
212. The Exclusive Training of a Single Faculty impossible, 196.
213. The Sensations—How Trained, 197.
214. Manual Training—The First Step in Education, 197.
215. The Hand then should be Trained, 199.
216. The Sense of Sight Trained by Exercises upon visible Objects judiciously selected and arranged, 199.
217. The Sense of Hearing Trained by listening to significant and musical Sounds, 201.
218. The Training of Each Intellectual Sense stimulates the other two Senses to Disciplinary Action, 201.
219. The Memory Trained by what Means? 202.
220. Memory Trained by vivid Con- [cepts, 202.

SEC.
221. Exercises that Train the Senses —Train the Memory also, 203.
222. Interest and Reiteration, 203.
223. The Perfect Classification of Concepts help their Retention in Memory, 204.
224. Conception how Educated, 204.
225. The Concepts and Processes that train the Faculty of Analysis, 205.
226. The Training of Abstraction— How Conducted, 206.
227. Ways and Means of Educating Imagination, 207.
228. The Special Method of Training Imagination, 208.
229. The Classifying Faculty—How Educated, 209.
230. The Concepts and Processes that Educate Class Judgment, 210.
231. The Means of Educating the Reasoning Faculty, 211.

CHAP. XVII.—*Expression as a Means of Intellectual Discipline.*

SEC.
232. Expression, 214.
233. The Means of Expression in Natural Language, 214.
234. Character of the Sounds in Natural Language, 215.
235. Judicious Training in Expression by Means of Natural Language, 215.
236. The Means of Expression in Artificial Language, 215.
237. Artificial Language Addressed to the Ear and the Eye, 216.
238. The Instruments of Expression in Artificial Language, 217.

SEC.
239. The Value of Spoken and Written Language as Gymnastics for the Intellect, 217.
240. Effect of Expression on the Concept further Explained, 218.
241. Every Faculty has a Language which is essential to the Clearness of its Products, 219.
242. The Comparative Importance of Language in Courses of Study, 221.

CHAP. XVIII.—*Higher Spontaneities Springing from Trained Effort.*

SEC.
243. The Higher Spontaneities how produced, 223.
244. The Will Directs the Higher Spontaneities in Groups, 224.
245. Illustration of the above principle, 224.

SEC.
246. Example from Reading and Classification, 225. [strated, 227.
247. The Value of Reviews illu-
248. Conclusions reached, 228.
249. Maxims derived from the above Facts, 229.

CHAP. XIX.—*Injurious Effect of Wrong Arrangement of Studies.*

SEC.
250. Adjustment of Objects to Faculties, 232.
251. Violation of this Principle, 233.
252. Untimely Effort a Serious Obstacle, 235.

SEC.
253. Conventional Forms when learned, 236.
254. What the Serial Unfolding demands, 236.

CHAP. XX.—*Studies must be Selected that will Discipline the Faculties Strictly in the Order of their Development.*

SEC.
255. The Sciences as Gymnastics, 238.
256. Illustrated by Systematic Botany, 239.
257. Zoology submitted to the same Test, 240.
258. Mineralogy also, 240.
259. The Mathematical Sciences—Arithmetic its Gymnastic Value, 242.
260. Algebra, 244.
261. Geometry, 245.

SEC.
262. Language, Names, Reading, Writing, 246.
263. Composition, 248.
264. English Grammar, 249.
265. Rhetoric, 250.
266. Geography, 251.
267. Table showing the earlier mental operations in Concrete Numbers, 253. [254.
268. Classification of the Sciences,
269. Table Suggesting Partial Series of Studies, 255.

CHAP. XXI.—*Arrangement of Studies and Method of Instructing in Early Educating.*

SEC.
270. The Period when Formal Education should begin, 258.
271. The Age when Incipient, Systematic Effort begins, 259.
272. Value of Play-schools, 259.
273. Objects for the Formal Training of the Hand, 260.
274. Objects that Train the Sense of Sight exclusively, 261.
275. The Natural Order of Presenting Colors to the Eye, 262.
276. The Tints and Shades, 263.
277. The Tinges, 264.
278. Formula for Producing the Compound Colors, 267.
279. Objects used for the Simultaneous Training of the Hand and Eye, 267.
280. The Forms of the Triangle, 268.
281. Equipment, 268.
282. Manual and Visual Exercises on Triangular Forms, 269.
283. Other Plane Figures, 271.
284. The First Squares and Parallelograms, 272.

SEC.
285. Added Suggestions, 272.
286. The Squares drawn on the three Sides of a right-angled Triangle and the Circle, 273.
287. The Simultaneous Training of the Ear and the Tongue, 274.
288. The Sounds Best Adapted to Training the Ear and the Tongue, 275.
289. Musical Sounds, 275.
290. The Elementary Articulate Sounds, 276.
291. Training the Eye, Ear, Hand, and Tongue by naming the Capital Letters, 277.
292. The Training in Concrete Numbers Preparatory to the study of Arithmetic, 278.
293. Lessons on Irregular Forms in Botany and Zoology in connection with Drawing, 279.
294. Lessons on Forms in Zoology, 280.

The Teacher's Psychology.

Chapter I.

INTRODUCTION.

TERMS DEFINED AND THEIR MEANINGS ILLUSTRATED.

1. The Value of Definite Terms.—To the student of psychology a clear and definite knowledge of the nomenclature which the science employs, is of the utmost importance. Indeed, it is quite impossible that terms whose significance is not completely grasped, should guide us in the close self-scrutiny that discriminates and classifies our mental operations.

It is, moreover, essential to genuine progress in studying psychology, that every term should be used for no other purpose than to designate, with unvarying exactness, a single mental element. One of the sources of confusion which the young student has hitherto encountered on the very threshold of the science, lies in a nomenclature which contains many words of varying significance. For example, the term *perception* sometimes means the POWER of perceiving, sometimes the *act* of perceiving, and frequently the *idea* which the act of perceiving produces in the mind.

Let us at the outset remove this ancient hindrance

to a fascinating study, (1) by learning with precision the exact meaning of every psychological term which the science employs, (2) by restricting every term to the designation of an identical object, (3) by realizing in our own thought the mental element which each term specifies.

2. Mind is the name of an indivisible spiritual force that manifests itself in thinking, feeling, and willing. I learn that my friend has secured an honorable appointment; I rejoice in his good fortune, and determine to tell him so. Here the first sentence, " I learn that," etc., expresses an act of knowing; the second, " I rejoice in his good fortune," embodies a feeling—that of joy; while the third, "I determine to tell him so," affirms an act of the will.

3. Knowing is a term which denotes the act wherein the mind affirms the certainty of that to which it attends For example, I witnessed yesterday the burning of my neighbor's house. *I knew* through the evidence of sight and hearing that the dwelling was being consumed by fire. To-day *I know* through memory the event which left my neighbor homeless. Thus the word *knowing* signifies the act in which the mind affirms the facts which it gathers from experience or from its own operations.

4. Feeling.—Feeling is the term which expresses the pleasant or painful states of mind which are produced each by the presence of its peculiar cause. For instance, the sting of a wasp produces a feeling of bodily pain. Personal triumph elicits a feeling of joy; the death of a friend begets a feeling of grief. The presence of peril causes fear. Outrage arouses anger; favors voluntarily bestowed excite gratitude. Sympathy and kindness received from others elicit our love. Now joy, gratitude, and love are manifestly pleasant feelings; while grief, fear, and anger are painful ones. Thus every feeling to which we are susceptible is either pleasant or painful. Moreover, in feeling the mind is manifestly *passive*, while in knowing the mind is *active*.

The wish or longing for gratification, which naturally attends a feeling of whatever class, is called a *desire*.

5. **Willing** is a word which signifies the choice the mind makes of the desires it will strive to gratify. Thus I desire to visit the city on urgent business, but a storm is impending, and I desire to avoid exposure to it. Which desire shall I gratify? In consideration of safety to health, I decide to remain at home. This act of choice between two opposite desires is expressed by the term *willing*, and the desire which the mind decides to gratify is its *motive*.

6. **The Will—Effort-making.**—Further, the power which the mind puts forth in the act of willing is called the *will*. But the will is not only the power of choosing between two or more desires as to which shall constitute its motive, but it is the impelling force which directs and sustains the action that follows. When I determine to walk to the station instead of riding, it is my will that not only makes the choice, but directs and impels, to its allotted purpose, every faculty of mind and every muscle employed in the complex act of walking. The acts of will are called *volitions*.

7. **Spontaneity — Self-movement.**—Any act of mind, which, being excited by the presence of an object, takes place without any effort of the will, is termed a *spontaneity* or a *spontaneous action*. Many of our actions are pure spontaneities, especially in their commencement. Each of the articles in this room, for instance, when in line with the open eye, produces

a spontaneous act of vision. I see the table, the sofa, the painting on the wall, by no effort of my will. But if I prolong this act of seeing for any purpose, however slight, it becomes a voluntary effort—in other words, an act of will All our mental acts *begin* as spontaneities, which are usually changed into efforts by an impulse of the will. The feelings, being incited solely by their peculiar objects, are spontaneous throughout.

8. What is a Faculty?—The word *faculty* signifies power. A mental faculty is the power which the mind possesses of acting upon any object, whether external or internal, in order to discriminate it from surrounding objects. My faculty of sight acts, let us say, upon a rose, distinguishes it from other flowers, and perceives it to be a rose. My faculty of conception acts upon the notion of Brooklyn Bridge, which memory recalls, and distinguishes it from the memory of all other bridges.

9. The Object, Action, and Product of a Faculty.—That on which mind acts in the exercise of any faculty, is termed the *object* of that faculty. I hear a peal of distant thunder; the sound so heard is the object of the faculty of hearing. The *action* of this faculty, is called *listening;* the product of such action on the object referred to, is a notion or idea of the sound of thunder. Further, if I examine a flower—say the violet—for the purpose of classifying it, it becomes the object not only of sight, but of another faculty of mind, namely, the power of analysis. In this case, I repeat, the faculty is that of analysis; the object is the violet, the action is *analyzing*, and the resulting product is a knowledge or notion of the parts and properties of the violet. Thus every faculty of mind has its legitimate objects, acts, and products, each of which in the study of psychology should be carefully discriminated from the other.

10. Attention — The Centering of Effort.—Attention is the centering of the act of any faculty upon its object, by an impulse of the will. In an art-gallery I notice spontaneously a group of paintings. By an effort of the will I may concentrate this spontaneous act of vision upon a single figure, inspecting its outlines, coloring, execution, etc. This act of narrowing the range of involuntary vision and fixing it by an exertion of will on a single object, is called *attention*. Attention concentrates any mental act upon the object under scrutiny and excludes adjacent objects. It is the essential factor in all mental exercise.

11. What is a Sensation?—*Sensation* is the term that designates the local feeling which is produced by the contact of any organ of the senses with its peculiar external object. The odor of a heliotrope when held near the nose begets the sensation of smell. The pulp of an orange in contact with the tongue, produces the sensation of taste. The prick of a pin incites the sensation of feeling. All the sensations, like other feelings, are either pleasant or painful. When pleasant, they generally attract us to things that are healthful to the body; when painful, they warn us against things that are harmful.

12. Sense-perception—The Gathering of Percepts.—*Perception* is the general name of a faculty through whose action the mind gains knowledge, whether of things without or within ourselves. Sense-perception is the faculty which supplies the mind with knowledge of external objects through the action of the senses of touch, sight, and hearing. In exposure to a storm I see, hear, and feel the driving rain. In this act of sense-perception the senses employed are those of touch, sight, and hearing. The object is the driving rain; the acts put forth are feeling, seeing, and hearing; and the product of these acts, while in progress, is a notion or *percept* of the rain. The percept in this case unites in itself the elements gained from feeling, seeing, and hearing. If the object of my sense-perception had been a thing which was visible but not tangible or audible, as a picture, a cloud, or a rainbow, the percept would have contained only the elements gained from the act of sight.

13. A Product, then, in terms of psychology, is the knowledge, notion, or idea which results from the action of any faculty upon its object; and a *percept* is the immediate product of sense-perception when acting on an external object through one or all of the three senses named.

The four aspects of sense-perception, which for the sake of accuracy I have discriminated by different names, may be indicated in consecutive order, as follows:

Faculty.	Object.	Action.	Product.
Sense-perception.	External Thing.	Touching. Seeing. Hearing.	Percept.

14. Memory—Storing Concepts.—Memory is the faculty which unconsciously receives, retains, and restores the products or ideas gained through the action of the other faculties. For example, my memory at this moment restores the idea or notion of the Cathedral of Strasburg, which I once visited and examined, thus attaining an idea or percept of its structure, form, size, style, etc. This percept, which I acquired through the sense of sight, was stored and retained unconsciously in memory, and is now recalled therefrom and consciously recognized by my mind as the idea or concept of the Strasburg Cathedral. It is manifest that all knowledge gained from whatever source, except what is before the mind at the present moment, is held unconsciously in the memory.

Arranging the operations of memory in its relation to the senses, we have the simple synopsis:

Faculty.	Object.	Action.	Product.
		Receiving.	Idea
Memory.	Percept.	Retaining.	or
		Recalling.	Concept.

15. Conception—Holding Concepts.—Conception is the faculty that realizes and holds consciously before the mind an idea or concept recalled from memory. Thus when memory restores the idea of the Strasburg Cathedral, my faculty of conception instantly grasps, recognizes, and holds up this idea for conscious contemplation. In other words, the mind's eye sees the picture of the Cathedral once photographed on the memory and now recalled from it. This act of conceiving the idea of an individual object which the senses have gathered and memory retains and restores, is, I repeat, the conceptive faculty. The *object* of this faculty is any idea which memory furnishes, its action on this object is called *conceiving*, and its product is properly a concept which represents to the mind completely and vividly the material things on which the senses have previously acted.

A concept, then, is the product of the conceptive faculty, acting on any idea which another faculty has supplied and memory retains. The concept of the cathedral, being like its outside object composed of parts, is called *a concrete concept.* A synopsis of these elements of conception runs as follows:

Faculty.	Object.	Action.	Product.
Conception.	Unconscious concept of memory.	Conceiving.	Conscious concept.

16. Analysis—Dividing Concepts.—The faculty of analysis is the mental power which notes or scrutinizes, one by one, the parts and properties of which a concrete concept is composed. Let each of us now conceive of the dwelling in which our childhood was spent. This familiar mental picture or concept of our early home was gained, as a percept, by innumerable acts of sense-perception. It has, let us say, been since retained in memory, and, as now recalled therefrom, is held before the mind as a distinct, clean-cut concept of the house we first lived in. Fastening our attention persistently on this concept, what further operation does the mind begin instinctively to perform upon it? From contemplating the picture as a whole, do we not inevitably proceed to note its parts and properties; its color, form, size; its windows, its roof, its protruding chimneys? This is the invariable act of analysis that follows the act of conceiving a concrete object as a whole, and its product contains various concepts of the elements which the object comprises. The synopsis of the operation is as follows:

Faculty.	Object.	Action.	Product.
Analysis.	Concept.	Analyzing.	Concepts of properties and parts.

17. Abstraction—Transforming Concepts.—The faculty of abstraction is the power that, by comparing each property previously revealed by analysis with the same property found in other concrete objects, gains therefrom an *abstract concept*. An abstract concept is a notion or idea of any property apart from the concrete things in which it exists. Recall the concepts of our early home, and review the properties disclosed by the act of analysis. Inspect specially the color which the mental picture presents, and suppose it to be white. Now, how am I able instantly to recognize and name this color white? Simply because, in my earliest mental experience, I compared my notion of white as found in one object, with my notion of white as found in another, and perceiving them to be identical, finally attained the notion of white apart from the particular object to which it belongs. I compared, for example, the white of the house I lived in with the white of a neighboring church, the white of snow and so on until I gained an idea of the color white outside of the special thing characterized by it. This idea is the abstract concept of the color under inspection, the process by which it is acquired is called *abstracting*, and the objects thus transformed by the process are the concepts of individual properties revealed by analyzing the concrete concepts. The synopsis of abstraction is as follows:

Faculty.	Object.	Action.	Product.
Abstraction.	Individual concepts of properties.	Abstracting.	Abstract concept.

18. Imagination—Building Concepts.—Imagination is the name of the faculty which constructs at pleasure new concepts out of the material supplied by the three preceding faculties. To the concept of the house wherein I was born and bred, I added from time to time, through sense-perception, the concepts of other houses. These concepts, under the operation of analysis, furnished my mind with numerous concepts of their parts and properties, each followed by a corresponding abstract concept. In consequence, my memory is stored with (1) concrete concepts of the houses observed ; (2) concepts of their particular parts, as doors, windows, roof, chimneys ; and concepts of their properties, as size, height, contour, color. From these last and similar concepts of individual objects I have evolved by comparison the corresponding abstract concepts of size, height, contour, color, etc. Now by modifying and combining these various concepts, my mind is able to construct a mental image of a house which is strikingly different from any I have ever seen. This new mental product is called *an image concept*, and the faculty producing it from the materials named is imagination.

SYNOPSIS.

Faculty.	Object.	Action.	Product.
Imagination.	Concrete concepts. Concepts of properties and parts. Abstract concepts.	Imagining	Image-concepts.

19. Classification—Grouping Concepts.—Classification, or the classifying faculty, is the power which arranges the individual concrete concepts gathered by the senses in groups or classes, on the basis of their resembling characteristics. We have seen that the mind gathers through the senses individual concepts— for instance, of houses ; that the analysis reveals, one by one, their parts and properties ; and that each property is transformed primarily by comparison, into a corresponding abstract concept. Guided by these abstract concepts, the mind scrutinizes successively the characteristics (properties, structures, purpose) of each house ; and finding that they resemble each other, arranges the objects to which they belong in a single class under the common name *house*. This concept of a class of things joined together by one or more common characteristics and designated by a single name, is called a *class-concept*. The faculty which thus groups single individuals into classes, is the classifying faculty, and the objects on which it acts are individual concepts that have resembling characteristics.

SYNOPSIS.

Faculty.	Object.	Action.	Product.
Classification	Concrete concepts. Concepts of parts and Properties. Abstract concepts.	Classifying.	Class-concept.

20. Judgment — Connecting Concepts. — Judgment is the faculty which affirms that an individual is contained in a class, or that a narrower class is contained in a wider one. In the sentence "Thomas is a soldier" I affirm that Thomas is included in the class denoted by the word *soldier*. In the sentence "A rose is a flower," I affirm that the narrower class termed *rose* is included in the wider class termed *flower*. In other words, roses as a class form a part of the wider class, flower. The sentences, "Thomas is a soldier," "A rose is a flower," each of which expresses an act of judgment, are called *propositions*.

SYNOPSIS.

Faculty.	Object.	Action.	Product.
Judgment.	Class concept.	Judging.	Affirmed concept.

21. Reasoning—Deriving Concepts.—Reasoning is the faculty that derives new truths or concepts from class concepts already known. Thus, to take the simplest example of reasoning, if I see an individual mocking-bird whose voice I have never heard, I infer nevertheless that he is a singer, because I know that the class of birds to which he belongs are singers. This operation of the reasoning faculty may be formally expressed as follows:

>Mocking-birds are singers;
>This is a mocking-bird;
>Consequently, it is a singer.

Here our object is the class-concept affirmed by judgment in the first proposition, and our product is the conclusion or inferred concept expressed in the last proposition.

SYNOPSIS.

Faculty.	Object.	Action.	Product.
Reasoning.	Concept of judgment.	Inferring.	Inferred concepts.

QUESTIONS ON CHAPTER I.

What is the value of definite terms in Psychology? When are terms the source of confusion? How can we remove the obstacles to successful study which terms of varying significance present? Meaning of the word *mind*. Illustrate. Significance of the word *knowing*. Meaning of the word *feeling*. Also of the word *willing*. Illustrate the acts of knowing, feeling, and willing. Define and illustrate the term *will*. What is a spontaneity? Define and illustrate. What is a faculty? Give the distinctions between the object, action, and product of a faculty, and illustrate. What constitutes the act of attention? Define and illustrate a sensation. Sense-perception, its nature and purpose; its object, action, and product. Define and illustrate memory. Define and illustrate conception, distinguishing its object, action, and product. What is a concept? Define analysis, discriminating its object, action, and products. What is the province of abstraction, and what are its products? Explain the process by which abstract concepts are gained. What is the peculiar office of imagination, and from what sources does it gain its materials? Give the order of its objects, action, and product. What is the special province of classification? Explain its operation, and give its object, action, and product in order. Explain the office, action, and products of judgment. What is the name of a sentence which expresses judgment? Define reasoning, and explain the process by which it derives new truths. Review and repeat from memory the synopses of the object, action, and product of each faculty.

Chapter II.

MIND AND ITS THREE MANIFESTATIONS.

22. Only Two Forms of Existence.—In our present state only two forms of existence come within the range of our knowledge. One of these is MATTER, the other MIND. We are not so constituted as to know either of these in its essence or inmost nature. Of neither matter nor mind are we capable of apprehending what it is in itself. We can only receive its manifestations and infer from these the reality of its existence. The manifestations or qualities of matter are its length, breadth and thickness, size, shape, roughness, hardness, etc. The manifestations of mind are its various activities, such as perceiving, remembering, feeling, desiring, etc. To our apprehension, then, matter is that which occupies space, resists pressure, impedes motion, is colored, figured, hot or cold; while mind is that which recognizes, recollects, imagines, reasons, is glad or sad, selfish or sympathizing. Apart from its qualities, I know of matter absolutely nothing; and apart from its operations, I know likewise absolutely nothing of mind. Through my senses I perceive the qualities of the one, and through my consciousness the operations of the other; but the existence of qualities necessitates the existence of a substance to which they belong. Shape, size, color, must be the shape, size, and color of something which is called *matter*. In like manner, actions compel the existence of an agent which puts them forth. There can be no act without an actor. Listening, judging, deciding, must be the efforts of something which listens, judges, decides, and which we call *mind*.

Hence, I repeat, we know matter only indirectly through its properties which are presented to the senses, and we know mind only indirectly through its operations which consciousness affirms. And finally, let us note that, as entities, we know as much of one as of the other, and that of neither do we know anything except its bare existence.

23. The Manifestations of Mind of Three Kinds. —Having found that every mind perceives directly its own operations, let us add that these operations are all included in three great classes. These classes are clearly expressed and discriminated by the participles, knowing, feeling, willing. Every movement which the mind originates, is either a knowledge, a feeling, or a volition. Our ordinary language accurately distinguishes these three kinds of mental phenomena from each other, and the most ignorant person appreciates at once the difference between them.

24. Knowledge.—Thus all the operations of knowing are expressed by active verbs. In the examples, "I see an object," "hear a sound," "remember an event," "infer a fact," each verb, with its object, designates a particular act of the knowing power. I simply affirm, in these several propositions, that I know something through sight, hearing, memory, and reasoning.

25. The Feelings.—On the other hand, the feelings are not activities, but states. Fear, anger, love, joy, for instance, are feelings, each of which occupies the mind as an effect due to the presence of its appropriate object. Thus impending danger excites fear, amiable qualities beget love, personal insult awakens anger, strange events produce surprise. Feelings, then, are different mental conditions, each of which is produced by the immediate influence of a particular cause. As simple states of mind, they are expressed by adjectives used as the predicates of neuter verbs. In the sentences " He is angry," " James is sad," " You are afraid," " I am joyful," the existence of these states is affirmed without reference to their causes. When, however, we desire to express the feelings *as effects*, we employ as predicate the perfect participle. Thus, terrified, pleased, gladdened, shamed, encouraged, signify not only states of mind, but suggest the causes that produce them.

26. Knowledge Precedes Feeling.—It is manifest that knowledge is the invariable antecedent of feeling. In every instance we must know an object before it can affect our sensibilities. I must perceive danger before I can experience fear ; I must appreciate an offence before I can feel resentment. I am not conscious of love until I appreciate the traits of character that call it forth. Feelings, whether they are passions or emotions, are uniformly produced by the *known* qualities or characteristics of objects which are fitted to inspire them. These objects, as the causes of the feelings, may properly serve as the bases of their classification.

27. Sensations.—A sensation is a feeling caused by contact of a bodily organ with its appropriate external object. Thus the odor of eau-de-Cologne in contact with the nerves of the nose, begets the sensation of smell. Food of whatever sort, when masticated and moistened by the saliva of the tongue, produces a sensation of taste. Contact of the fingers, or indeed of any part of the body, with the surface of a solid causes a sensation of touch. Now these and all other sensations, which are felt invariably in the organ wherein they are produced, vary in their intensity from the slightest to the most acute. The taste of milk, for instance, is mild compared with the taste of vinegar, and the sensation which springs from the handling of a thistle, has more strength than that which arises from the touch of velvet.

28. Sensation either Pleasant or Painful.—The sensations, like all the other feelings, are either pleasant or painful. The pleasant sensations indicate to the mind that the objects which cause them are healthful and useful to the body, and consequently attractive. The painful sensations announce to the mind that the objects causing them are deleterious to the body. Thus the sensations, since they enable us to choose with unvarying certainty the things that are helpful to the body, and to reject the things that are harmful, are the natural safeguards of our physical organization.* In the order of our experience sensation precedes knowledge, and is, consequently, an exception to the general statement that knowledge precedes feeling. The sensations are produced by physical contact and not by an idea of their causes.

29. The Appetites.—The desires we feel to satisfy the wants of the body, as indicated by the sensations, are called *appetites*. Among the more conspicuous of our appetites are our desires for food and drink, termed *hunger* and *thirst*. But the desires we have for rest when weary, for sleep when drowsy, for warmth and comfort and relief from pain, are, with equal propriety, denominated *appetites*. In short, any bodily desire that prompts us to awaken the pleasant sensations which contribute to the preservation of the body, or to the perpetuation of the race, is an appetite. It is manifest that an appetite is strictly a desire to produce and gratify a sensation.

---

* Sensation is more fully explained in Chapter III.

The appetites are susceptible of different degrees of intensity, which vary with the urgency of the bodily wants they supply. My desire for food supplied in abundance at regular meals, is comparatively moderate. Reduce the food to a quantity below the needs of the stomach, and my desire becomes a longing. Deprive me of food altogether, and the longing becomes a craving. Thus we perceive the perfection of the wonderful adjustment by which the sensations and the appetites minister to the wants of the body.

When, as in infancy, the appetites seek for gratification without the guidance of a deliberate purpose, they are termed *instincts*.

30. Selfish Feelings—Egoism.—Besides the sensations and the bodily appetites, which are in their nature purely selfish, I have other feelings whose primitive purpose is to secure the satisfaction, safety, and well-being of myself exclusively. Whenever danger of any sort threatens injury to my person or to any closely related interest of mine, I instantly experience the feeling of *fear*. When assailed with insult, calumny, or any form of outrage, I am at once conscious of a feeling of anger. From these feelings of fear and anger, whose purpose is self-defence and self-preservation, spring two opposite impulses. The one prompts me to escape the menacing object; the other to attack the offender, and return injury for injury. Fear, if often experienced, begets a kindred feeling more permanent and less intense, namely, *humility*. Anger repeated naturally results in *hatred* of the person causing it. Malevolence is a dominant feeling of hostility to those around us. Envy, jealousy, spite, and many similar states of mind are in the same category of feelings that are exclusively selfish.

Scrutinizing closely the character of these feelings,

we find (1) that they form a closely related cluster, (2) that they are uniformly painful, (3) that in excess they are brutal, and (4) that one of the purposes of a true education is to secure their habitual subjection to the higher feelings.

31. Selfish Feelings that are Pleasant.—But I am susceptible, as a human being, to other self-centering feelings than the painful group just mentioned. I am capable of self-love, self-esteem, self-approbation, pride, etc. Self-love, if immoderate, annihilates sympathy for others; self-esteem, when excessive, becomes self-conceit; self-approbation in the extreme, culminates in self-homage ; and pride, when it dwells with complacence on merits that are trifling, degenerates to vanity. I also have a capacity for love of possession, love of power, which, when sought for my own good solely, are also selfish. These, and all other feelings whose end is our personal welfare, accomplish their legitimate purpose when held habitually in check by the promptings of duty and right.

Manifestly the selfish feelings are, like the appetites, subject to varying degrees of intensity, due both to their sensitiveness and to the character of the objects that call them forth. Anger rises sometimes to rage, even to uncontrollable fury. Fear, under favoring conditions, reaches a state of mind expressed by terror or dismay. Malevolence, if long indulged, becomes enduring malignity. It is clear that human beings are susceptible of the selfish feelings in common with the lower animals.

32. The Social Feelings—Altruism.—As the selfish feelings are based on our love of self, so are the social feelings based upon the love we cherish for our fellow-men. This is the class of feelings whose growth made it possible for civilization to emerge from bar-

barism. It is his capacity for sympathy with others that elevates man above the brute, and enables him to organize society and to maintain its institutions and its laws.

Let us note briefly the characteristics of the social feelings that are most conspicuous among men. An enduring love for those around us leads us to the practice of self-denial and self-sacrifice for others' good. Sympathy, as the term implies, participates spontaneously in the joys and sorrows of our neighbor, and makes them our own. Pity commiserates suffering wherever found, and naturally seeks to give it relief. Gratitude is an appreciation of the favors freely accorded us by others, and a sense of the obligation incurred thereby. Generosity is the feeling which prompts us to bestow without stint, upon those around us, anything we possess which will increase their comfort or happiness. Patriotism is the feeling of attachment we cherish for our country, its institutions, laws, and the body of its citizens. Philanthropy gathers within its comprehensive scope the entire human race, and ardently desires the advancement and happiness of all mankind. The above constitutes a group of closely related social feelings, which generally coexist in the mind they characterize. It is evident that society advances in the scale of civilization in proportion as the social feelings prevail in human character over the selfish feelings.

33. The Emotions.—Still higher in the scale of human character are the feelings termed the *emotions*. It is susceptibility to the emotions that distinguishes man as a being superior to the lower animals. The germs of the higher feelings are occasionally exhibited by an animal that is unusually intelligent, but they are uniformly incapable of further development. In the human mind alone their expansion under the influence of culture seems unlimited.

34. The Emotion of Beauty.—If my senses of sight and smell are affected by a heap of decaying rubbish, I am at once repelled by a feeling of disgust; but when my vision gathers in a cultivated landscape wherein bright colors and graceful forms are harmoniously blended, I am instantly conscious of a serene pleasure. This is the emotion of beauty. There is a marked difference in the susceptibility of individual minds to this elevated feeling which the perception of beauty produces. The man whose taste is cultivated discerns unerringly all the manifestations of beauty in nature, in art, and in human conduct. The man whose taste is uneducated, and consequently obtuse, is blind to the revelations of beauty, except in their simplest aspects.

To conceive and express beauty through colors, as in painting; through forms, as in sculpture; through sounds, as in music; and through metrical language, as in poetry,—is the province of *art*. To judge of the fitness in the products of art or nature to excite the emotion of beauty, is the province of *good taste*. Manifestly the mental refinement to which we attain, depends on the special culture that improves the taste and quickens the æsthetic sensibility.

35. Emotion of Sublimity.—The emotion of sublimity, which is more absorbing, and consequently more transient, than that of beauty, is awakened by whatever is grand or elevated, whether in the magnitudes of matter or in the achievements of mind. The contemplation, for instance, of a storm on the mountains, where,

" Far along,
From peak to peak, the rattling crags among
Leaps the live thunder,"

appeals forcibly to the emotion of sublimity; but when we reflect rightly upon the divine magnanimity, which freely sacrificed life to redeem the world, the same emotion rises to a loftier height.

36. Love of Knowledge.—One of the higher emotions of which we are capable, is the feeling of pleasure derived from the attainment of knowledge. In childhood and in conditions of mind akin to it, this feeling centres upon individual facts and events within the range of the senses. It is at this period that the simple love of novelties, which is termed *curiosity*, impels the child to examine and learn the properties of the concrete things around him. But in later life, the mind, which is incited to further study by the knowledge already acquired, finds a constant and refined enjoyment in the mental efforts that reveal in every fact a principle and in every event a law: thus the acquirement of knowledge in any department of science, yields to the earnest student pleasures that are elevated and unfailing—pleasures that are in marked contrast with the lower gratifications of animal life. These pleasures, which spring from intellectual activity rightly directed, constitute mainly the incentives to strenuous, persistent efforts of study that end in discipline. The interest felt by the pupil is the teacher's most effectual help.

37. The Moral Sense—Conscience.—The emotions of conscience, or the moral sense, are awakened in the mind by the perception of that which is right in our relations to men and to God. I owe my neighbor, for example, a sum of money which I promised to pay on or before a certain date. The fulfilment of this promise is not only a legal obligation, but is a duty. To neglect the payment, as agreed upon, would do violence to my sense of right. In all the transactions of men with each other, prompted by the selfish and the social feelings, there is a higher feeling caused by the knowledge that one act is right and another wrong. It is the feeling which impels us to do the right and shun the wrong. Between the selfish and the social feelings there is a frequent conflict, which can be settled only by the question whether the diverse actions they prompt are right or wrong. This question it is the province of conscience to decide, and to obey its dictates is our uniform duty.

38. The Religious Emotion.—Our feelings of reverence for and dependence on a supreme Being, whose existence we intuitively recognize, constitute the religious emotion. To feel the constant presence of an omnipotent Father, to whose infinite wisdom and goodness we are indebted for life and all its enjoyments; to cherish the impulses of submission and obedience to his laws, which such a feeling engenders,—these are the highest mental states of which we are capable. The love of God, with its attendant emotions that constitute worship, is the crowning capacity of the human soul.

39. Every Feeling attended by Pleasure or Pain.—Each of the feelings, whether higher or lower, is either pleasant or painful. Some of these, as anger, fear, jealousy, envy, hatred, are intrinsically painful; others, and by far the larger number, are pleasant. But every pleasant feeling becomes painful when the conditions that produced it are reversed. Thus let the person whose scholarship I admire, be guilty of falsehood or moral cowardice, and my admiration, which is pleasant, is changed to contempt, which is painful. Substitute deformity for beauty in any object on which I gaze, and the pleasant emotion of beauty would yield instantly to the painful emotion of disgust. When the moral sense prompts to duty, to obey begets self-approbation, which is pleasant; to disobey, self-condemnation, which is painful. Thus, in general, it is the province of pain to repel us from whatever does violence to our better nature, while, on the other hand, pleasure, especially in its higher aspect, attracts us to that which contributes to our well-being.

40. The Gratification of any Feeling uniformly Pleasant.—Whatever the character of a feeling, its gratification is uniformly pleasant. To fear, escape from danger is gladness; to anger, " revenge is sweet ;" to love, reciprocity and possession are a perpetual delight. This pleasure derived from the gratification of a feeling, is expressed by the general term *joy*, while the pain that results from a violence to feeling which prevents its gratification, is expressed by the general word *sorrow*.

41. The Desires.—Every feeling, of whatever class, is or may be attended by a longing for gratification. Thus fear begets a desire for safety; anger, for injury to its object; love, for the elevation, and envy for the humiliation, of its object. Patriotism incites a desire for the advancement of one's country; philanthropy, for that of the world. A desire may, like the feeling it follows, be a mild and transitory wish or an intense and permanent passion. In the former case, its influence on our conduct is comparatively slight; in the latter, it may give to life its purpose, and to character its strength.

42. The Will.—The will is the third division of our mental manifestations. Will is the power of choosing or deciding between the desires that the mind considers, which desire it will strive to gratify. It is not only the power of choice, but the force that directs and impels all the acts of which we are capable. It controls and guides every intellectual faculty to the attainment of its allotted purpose. It is the general propelling power without which the mind would be given over to its own spontaneities.

43. Will no Immediate Control over the Feelings.—But though the will, under the influence of desire, determines, directs, and quickens every process of thought, it has no immediate control over the feelings. No direct effort of the will, however strenuous, can give existence or duration to this or that state of mind. I cannot be happy or melancholy by simply willing it. By no mental exertion, can I feel admiration except in the presence of the qualities that naturally call it forth. But the will may indirectly produce any mental state by presenting to the mind its natural object or cause. We are unable, for instance, to feel an emotion of pity by simply striving to feel it. But if we think of an example of suffering, the emotion springs up spontaneously. Thus the will, while it maintains the immediate control of the knowing power in all its operations, has over the feelings only an indirect influence.

44. Will uniformly preceded by Desire.—There can be, then, no exertion of the will except in the presence of one or more desires. It is desire that invariably prompts to the act of choice. A man without desires could make no voluntary efforts. He would be like an automatic machine or like a rudderless ship on the ocean. But the will may stimulate or suppress our desires by the same principle that enables it to influence the other feelings. In other words, it can turn the attention of the mind from the object of a present desire, and fix it upon an object which tends to excite a different desire. Thus, under the impulse of appetite, a man may desire to drink brandy, but by thinking, through an effort of the will, upon its harmfulness, he may check this desire and choose to abstain from gratifying it.

45. Motive—Freedom of Choice.—Let us suppose that several desires of equal urgency are present to the mind at any given moment. The will is free to choose from these the desire which it will strive to satisfy. The desire so chosen becomes *the motive for action.* This freedom of choice among the present desires of that one which shall be its motive, may be easily illustrated. For example, I am exhausted and drowsy from long wakefulness, and desire to sleep. This is a bodily appetite. At the same time I have valuable property which is exposed to destruction from a threatening storm, and I desire to attend personally to its safety. This is a desire which springs from a selfish feeling. At this moment I hear that my neighbor has been injured by an accident, and I desire to give him such immediate relief as my presence can afford. This is a desire prompted by social feeling. A fourth desire, which in similar conditions is inevitably present to a well-regulated mind, is the wish to satisfy my *conscience.* This desire emanates from a higher emotion, namely, that of the moral sense. It is a desire to do right.

Now which of the acts that the three aforesaid desires urge me to do *is* right? Each of these desires is based upon its peculiar obligation. The first is an obligation to preserve my bodily health. The second is an obligation to secure my property from harm. The third is an obligation to help my neighbor in his distress. I can fulfil but one of these obligations. My settled habits and inclinations may give to one of them the greatest *urgency,* but which one under the circumstances has in *itself* the greatest weight? The answer is happily close at hand. The weight of that obligation is greatest whose fulfilment, in the emergency mentioned, conscience affirms to be right. My will thus influenced by conscience decides, let us hope, in favor of the desire which originates in sympathy; and forgetting the promptings of self-interest, I hasten to the relief of my neighbor. When, therefore, a desire to satisfy conscience harmonizes with and approves any one of the other desires, higher or lower, the will, if properly educated, makes that one the motive for an effort whose purpose is to gratify it. But it is, of course, clear that the will is free to reject the demands of conscience and yield to the urgency of a lower passion despite its immoral nature. And there is a constant tendency of the higher impulses to surrender to the lower, in minds whose passions are habitually importunate and whose moral sense is feeble.

46. The True Motive.—Beyond question, our appetites and self-seeking desires answer a legitimate purpose in the economy of a well-ordered life. The end they serve when judiciously gratified, is the well-being of one's self both in body and mind. The desires for food, rest, comfort, personal safety, and personal distinction become genuine motives for efforts of the will whenever their temperate gratification does no violence to the moral sense or the rights of others. And this is true in by far the greater number of instances. But if a selfish desire which the will accepts, is in collision with the dictates of duty,—as when one longs to get possession of his neighbor's property by unlawful means,—the motive based upon it is *false*, and the moral sense is blunted thereby. Thus, in every conflict of desires, conscience is the supreme arbiter in settling which will constitute the true motive for action, and the right discipline of conscience is therefore a matter of vital moment.

47. High Moral Character.—The attainment of high moral character is the leading purpose of every true and noble life. Such a character consists (1) in an infallible judgment of the distinction between right and wrong in human actions, (2) in the settled predominance of the higher desires over the lower, (3) in the habitual choice by the will of those desires, as motives for efforts, which harmonize with the dictates of a cultured conscience. A high moral character is, consequently, the product of educated moral perceptions, of life-long self-denial of animal passions, of sympathies which are uniformly sensitive to the rights of others, and of a will trained by persistent practice, to select infallibly for its motives, the desires which accord with impulses of the moral sense.

48. The Order of Sequence.—The order of manifestation held by the sensibilities and the will is evidently as follows:

1. The cause of feeling as perceived by the mind. 2. The feeling as its effect. 3. Desire as the product of feeling. 4. The will selecting the desire as its motive for the effort it determines to put forth.

Feeling and desire are only two aspects of a single state of mind.

49. The Order of our Mental Operations.—The fact that knowledge is preliminary to feeling, and that feeling is the incitement to choice, produces an unvarying order in our mental operations. To know, to feel, to decide,—this is the life-long, uniform round. I perceive on the shelf of a dealer a new book: it awakens an interest and a desire to know its contents. I consequently decide to purchase it. I see dark clouds above the horizon, and hear peals of thunder: the fear of a drenching and a desire to escape it follow. Accordingly, I determine to seek the nearest shelter. I study a painting in the gallery of the Louvre: its artistic excellence excites an emotion of beauty and a longing to possess it; but I decide to forego its possession because of its great cost.

50. Rapidity of our Mental Operations.—Our mental movements are frequently exceedingly swift and subtle, and each of the threefold operations described often flashes through the mind in a twinkling. The rapidity of their occurrence is such that they coexist in consciousness at any given instant, yet in their commencement they follow without variation the order I have named.

51. Names which Designate the Triple Phenomena.—The three great classes of mental manifestations explained above, are designated in our daily language by uniform generic names. All the phenomena of knowing or thinking are included under the term *intellect* or *intelligence*. The entire range and variety of human feelings are gathered under the word the *sensibility*, or its plural, the *sensibilities;* while the will has several synonyms which express its *action*, such as decision, choice, determination, resolution. It should be remembered that mind itself is a *single* entity, an indivisible unit, and its various manifestations, which we call *experience*, constitute the triple phenomena through which mind is known. Thus:

Mind. { Intellect. Sensibilities. Will. }

52. Consciousness is the constant knowledge which the mind has of its own acts and operations. It is through consciousness that the mind is aware of all that passes within it. I cannot think without knowing that I think, or feel without knowing that I feel, or will without knowing that I will. In consciousness, the mind not only recognizes each one of its own movements, but distinguishes this movement from all others of like character. Thus in a fit of anger I know that I am angry; and further, I discriminate this passion now uppermost within me from all other passions of which I am capable. I am conscious that it is anger that I feel, and not fear or love or shame. So also when I look at any visible object— as a book, for instance—I am conscious of this special act of vision, as distinguished from all other intellectual acts, whether of sight, hearing, or memory. I know that at this moment it is a particular book I see, and that I see it to the exclusion of other books and all other objects of attention. Or further, if I recall the image of my deceased father, I am instantly aware that this is an act of recollection, and that the image before my mind is one that memory had previously stored, and not an actual external object revealed to the senses. Thus in our waking hours the mind is constantly conscious of all that goes on within it, recognizing distinctly every present act and state.

53. Consciousness, a Present Knowledge.—Further, consciousness is a recognition by the mind of that only which is actually before it at the present moment. I am conscious of each change in my thought or feeling while it is taking place. I cannot be conscious of that which is past or that which is to come. I am only conscious of the mental experience of the instant, and this consciousness continues and closes with the experience itself. For example, recalling the main street of a certain city to memory, I think in succession of the prominent objects contained therein. First a mental image of the street in general and then the images of the city hotel, the opera-house, the exchange and two churches, are reproduced and passed, each in its turn, under the eye of consciousness. Now consciousness dwells upon each one of these images until it disappears and the next one takes its place. That is, the consciousness which the mind has of any of its own acts, is exactly coextensive with the act itself.

54. Consciousness recognizes the Mind as the Originator of its own Acts.—But consciousness does not simply recognize and discriminate every act of its own while occurring; it recognizes the mind itself, as the author of such acts. Every sentence we utter, discloses this fact. When I say that I see a picture, that I remember a cyclone, or that I imagine a shipwreck, I am not merely conscious of the present mental acts which these sentences express, but I am conscious that I am the originator of these. In other words, the mind not only knows and discriminates the act of seeing a picture, recollecting a cyclone, imagining a shipwreck, but discerns the fact that each of these is its own act.

Now a sentence is a consciousness expressed in language, and, if we analyze psychologically any example of it, we shall find at least three mental manifestations.

1. That the mind is conscious of the act affirmed.
2. That the mind distinguishes this act as a particular kind of mental movement.
3. That the mind recognizes this act as its own.

55. Consciousness Identical with each Mental Act.—So far I have treated consciousness as a power separate from the mental act to which it testifies. In no other way, could its nature and office be set in the clearest light. Consciousness is not, however, in any respect different from the act to which it bears witness.

The consciousness of a thought and the thought itself are one and the same. The essence of every mental movement, is that it reveals to the mind that gives it birth, its own intensity, its kind and origin. Otherwise it would have no reality. Hence the consciousness of an act of mind is essential to its very existence. I cannot see this book without being conscious that I see it. I cannot hear a sound without knowing that I hear it, nor can I recall the face of an absent friend without being conscious of so doing. The acts of seeing, hearing and recollecting, are only so many kinds of consciousness. The sentence, "I remember an event," and "I am conscious that I remember an event," are only different methods of expressing the same mental act. Consciousness, therefore, is the general name for all our possible mental operations, each one of which has a particular name to designate its kind and character.

QUESTIONS ON CHAPTER II.

What are the two forms of existence to which our knowledge is limited? In what respect do we know each of these? What are the manifestations of matter in distinction from those of mind? Through what faculty do we know the operations of mind? Give the three kinds of operations manifested by mind. How is knowledge expressed? What are the feelings, and how caused? What parts of speech usually express the feelings? What words express the feelings as effects? Which is the antecedent in mental action, knowledge or feeling? How are feelings produced? What is the basis of a classification of the feelings? Define and illustrate a sensation. What are the localities of sensation. Illustrate the varying intensity of sensations. What is the purpose of a pleasant sensation? Also the purpose of a painful sensation? Give the order of time in which sensation stands to knowledge. Define the appetites, and name the more conspicuous. Illustrate the varying intensity of the appetites. What is the purpose of the appetites? When is an appetite termed an instinct? What are the characteristics and purpose of the selfish feelings? Illustrate their effects. What is the effect of a true education upon the selfish feelings? What selfish feelings are pleasant, and what painful? The effect of selfish feelings if immoderate. Under what conditions do the selfish feelings accomplish their legitimate purpose? Upon what general sensibility are the social feelings based? What effect have the social feelings upon human advancement? The characteristics of the various social feelings. What is the rank of the emotions in human character? The higher emotions in the lower animals undeveloped. What objects beget the emotion of beauty? Difference in individual minds of the susceptibility to beauty. The effect of education on this susceptibility. The province of the arts in expressing the beautiful. The province of taste in judging of the beautiful and its effect. What characteristics of mind or matter produce the emotion of sublimity? Describe the emotion derived from the attainment of knowledge. What is curiosity? The love of knowledge the basis of our intellectual progress. Pleasures derived from knowledge a stimulant to study. How are the emotions of conscience awakened in the mind? What is the rank of this feeling, and to what actions does it prompt us? Conflict of the lower and the higher feelings to be decided by conscience. Describe the religious emotions, and name their rank. Every feeling either pleasant or painful. Some feelings intrinsically painful. How does a pleasant feeling become a painful one? Give illustrations. What, in general, is the purpose of pain? What is the beneficent office of pleasure? Illustrate the character of gratification. Meaning of the terms, *joy* and *sorrow*. What is the nature of the desires, and how are they produced? Give illustrations. Define the will, and give its two offices. Relation of the will to the feelings. The spontaneous character of the feelings. By what feeling is the act of willing uniformly preceded? Is it possible to will without desire? How are the desires stimulated or suppressed indirectly by the will? When does desire become a motive? The will free to select this or that desire. Give the example for illustration. In a conflict of desires, which one should uniformly predominate? Give the distinction between the urgency of an obligation, and its intrinsic weight. In what minds do the lower desires predominate? What is the true motive? When does a lower desire furnish a true motive? When is the motive a false one? What feeling is the true arbiter in settling the conflict between the higher and lower desires? Give the leading purpose of every true and noble life. Of what qualities does a high moral character consist? The order of sequence in feelings and the will. What is the order of our mental operations? Illustrate the rapidity of our mental operations. Names which designate the triple phenomena of mind. Give the distinctions between the intellect, the sensibility, and the will. What are the characteristics of consciousness? Give illustrations. Illustrate the fact that consciousness is a present knowledge which the mind has of its own acts. Consciousness recognizes the mind as the author of its own acts. Illustrate. What three characteristics of consciousness does the analysis of a sentence show? Is consciousness different, in any respect, from the act to which it bears witness? Can there be any act of mind without consciousness? Illustrate.

Chapter III.

ON THE INTELLECT.

THE SENSES—GATHERING CONCEPTS.

56. The Purpose of the Senses.—The senses connect the human mind with the world around us. They serve two important purposes; the one to protect and nourish the body, the other to furnish the mind with a knowledge of external things. The sense of taste, for example, which acts by contact with its object, reveals the qualities of food and supplies the body with nutrition. The sense of sight, on the other hand, acts at a distance from its objects, and conveys to the intellect ideas of visible things without. Taste yields from the contact of the tongue and palate with a sapid substance, a distinct feeling of pleasure or pain, and a notion of something agreeable or disagreeable in the substance that causes it. Sight simply gives notions of outside objects, as colored and shaped—notions that are retained with distinct outlines in the memory.

57. The Organs of the Senses.—The bodily organs through which we gather nutrition, either for the body or the mind, are five in number, namely, the nose, the tongue, the hand, the ear, and the eye.

The internal cavities of the nose, are lined with a delicate mucous membrane in which terminate the olfactory nerves; and floating particles coming in contact with these nerves, occasion the sensation of smell. The upper surface of the tongue and the back part of the mouth, are covered by innumerable papillæ, within which lie the extremities of the gustatory nerves. These nerves, when excited by contact with food which is masticated and moistened with saliva, produce the feeling called *taste*. These two organs are used for other purposes than for smelling and tasting. The nasal cavities serve as air passages in breathing and the tongue and palate are prominent organs of speech.

In the hand, we have two senses acting through one organ. Two kinds of nerves, namely, nerves of motion and nerves of sensation connect the ends of the fingers with the brain. The first proceeding from the brain to the muscles of the body and to the fingers' ends, are called the *efferent* nerves, the second running from the surface of the body and from the fingers' ends to the brain, are called the *afferent* nerves. The efferent nerves, by contact of the hand with a resisting surface, give us a perception of the resistance and shape of external bodies, while the afferent nerves yield from the same contact a sensation of heat or cold, etc.

58. Sense-perception and Sensation Distinguished.—Sense-perception is the act by which we gain, through sight or hearing or touch, a knowledge of some object outside of ourselves. Thus I see the desk before me; I hear the sound of distant music; in a dark room, at night, I place my hand upon an object and learn, by its mode of resistance, that it is, for instance, a watch. Now each of these movements is an act by which we get the idea of an object in the external world. It is an act of gaining knowledge through the senses mentioned. It is therefore, a sense-perception and, being an act of gaining knowledge, it must be classed with the first great division of the mind explained above, namely, the knowing power or the intellect.

Sensation, on the other hand, is a feeling produced by smelling, tasting, or touching an appropriate substance. For example, I smell a rose and there follows a feeling of pleasure. I taste an apple and a feeling of pleasure also results which I refer to the tongue and palate. I place my hand upon some surface and gain thereby a feeling of heat or cold. These feelings caused by the senses named, when in contact with their objects, are sensations. They are not qualities of the external objects. They exist in the mind and, since they are feelings, they belong to the second great department of mind, namely, the sensibilities. Sense-perception I repeat, is the act of gaining an idea of something outside the mind, through sight, or hearing, or touch as resisting the efferent nerves. Sensation is the feeling which arises within the mind from contact, with its objects, of the organs of smell, taste, or touch as affecting afferent nerves.

It may be added here that neither a sensation nor a sense-perception is ever pure and single. Knowledge and feeling are always in conjunction; neither can exist separately. In every sensation of the lower senses, there is always a dim perception, and a slight sensation accompanies every perception of the higher senses.

59. Contrast of Sensation and Perception in Consciousness.—I place an odorous object, say an orange, near my nose. At once I become conscious of a sensation that is moderately pleasant. I remove the orange and the sensation immediately vanishes. I can neither retain nor describe it. Again I bring the juice of the orange in contact with my tongue and I am now conscious of another sensation more pleasant than the last. This sensation increases as the liquid approaches the back part of the mouth, and so leads to the act of swallowing. So soon as the saliva removes the liquid orange, the sensation disappears. It cannot survive the absence of the cause. That gone, I can neither describe nor retain it. Now smell and taste a decoction of wormwood, and the same condition and results follow, with one excption. The sensations are now unpleasant. Again, place the hand upon a marble surface; the afferent nerves are affected, and there arises a sensation of cold. Dip the fingers into warm water, and there follows the sensation of warmth. Reduce the temperature of the marble, by exposure, to zero, and raise that of the water to the boiling-point, and the sensations following similar contact, would, though opposite in character, both be painful.

Thus we reach, by trial, the characteristics that divide all feelings into two great classes. Sensations of touch, taste and smell, are, like all other states of the sensibilities, either pleasant or painful. Lay a tennis ball in the palm of the hand and close the fingers upon it. The resistance you encounter to a complete shutting of the hand give you, at once, knowledge of something external. This is not now a mere feeling produced on you like that of taste or smell, but a perception of an actual outside object which reveals to the mind, by resistance to the efferent nerves, that it is solid, hard, round, smooth; in short, a ball. This act of gaining an idea of an outside object through touch, is a perception; and the operation belongs consequently to the intellect. Further, a thrush sings in a neighboring tree; I hear the music of its voice and the sound so produced has the same characteristic of externality as found in the case of the ball. The sense of hearing announces the sound as existing in space and not in me. I have here an instance of genuine knowledge gained through the ear, of something in the world of sound, and the act is a perception through the sense of hearing.

Again, standing at the window I look out upon the landscape. A multitude of objects without are at once visible: trees, lawn, and dwellings in the distance, shrubs and flowers near at hand, present themselves to the eye. I can fix my attention upon a single shrub, tree or dwelling, or I can gather them all in at a glance as a single landscape. In either case, I obtain, by the act of seeing, a knowledge of outward things, things which have color and size and form. I am conscious, in this act, of no sensation that can be compared with that of smell or of taste. I use my eyes as instruments for gaining ideas of visible objects without. The act is a perception through the sense of sight. It is an act of knowledge and not a feeling, and belongs, therefore, to the knowing power or the intellect. Thus we find, by consulting consciousness, that we have six senses; namely, three that yield sensations, as smell, taste, and touch as distinguishing heat and cold; and three that are instruments of perception, namely, hearing, sight and touch as discerning resistance. Let us now compare the products of these two triplets as existing in the memory.

60. Sensations and Perceptions as Compared in Memory.—When we attempt to recall from memory the sensations produced in smelling, tasting, or touching a warm surface, we are conscious of a failure. We recollect distinctly the acts of smelling, tasting, and touching, but we can, by no means, bring back to mind the resulting sensations separately from the acts that produced them. I can remember clearly smelling a rose, tasting an orange, and feeling the cold marble; but the pure sensations of smell, taste and temperature arising therefrom, I can neither clearly recall nor describe. They do not furnish the memory with distinct notions; they are not consciously present as objects of thought, after the organs that gave them birth, have ceased to act. If, indeed, I repeat the act of smelling the rose, tasting the orange or touching the marble I recognize the same sensations as resuscitated, but they are not distinct, clean-cut entities of the memory. We recognize sensations when repeated, but we cannot restore them as simple furniture of the memory.

Turn now to the products of sight, hearing, and touch as a perceptive sense, and we find the case far different. The closing of the fingers upon it gave me a knowledge of the tennis ball; through the listening ear I heard the thrush's song; with the open eyes I gathered notions of various objects in the landscape. When these acts of perception have ceased, I recall the notions gained thereby, with the utmost distinctness. I am conscious of ideas in the memory representing individually the ball, the song, and the trees, wholly apart from the particular efforts of the senses by which I obtained them. Indeed, I may have wholly forgotten these special efforts from which they originated, and still retain under the eye of consciousness definite individual ideas which can be recalled at will from memory, compared in judgment, or combined in imagination.

61. Names of Sensations and Perceptions Compared.—The dimness of the sensations, as entities in the memory, will account for the paucity in language of terms with which to designate them. We have indeed, the qualifying words, *pleasant* and *painful*, which denote the two great divisions in which the sensations, in common with all other feelings of the mind, are classed. But beyond the adjectives, *sweet*, *sour, tart, bitter, spicy,* and a few similar terms, which apply alike to the sensations of taste and smell, we have no single words which serve as names for the vast variety of sensations supplied by these senses. The same is true of the sensations which spring from tactual impressions. The adjectives, *hot* and *cold,* with a few adverbs which qualify them, include nearly all the words which signify degrees in the wide range of sensations arising from temperature. Even the limited number of terms in use have double meanings, since they are made to signify, sometimes the sensation, and sometimes the unknown quality in matter which produces it. Thus we may say "the smell of the rose," "the taste of an apple," "the warmth of the fire," meaning at will either the sensation produced by these objects or the occult property they contain which is its cause. In the lack of special words in the language for designating special sensations, we are wont to refer them to the things without, by which they are called; thus "the smell of violets," "the taste of bread or of venison," "the heat of the steam-pipe," etc., etc.

Attend now to sense-perception in respect to the same matter, and the case is found to be quite different. Every idea gained from the outer world by sight, hearing or perceptive touch, is distinctly named.

In fact every object of knowledge thus acquired, has several names each of which assigns it to the higher or lower class to which it belongs. *Sound, voice, music, song, carol,* may each be properly used to denote the song of the thrush. *Tea-roses, roses, flowers, blossoms,* are terms which may be applied properly to the things of beauty that decorate my table. And thus it comes that no small fraction of language consists of names belonging to objects of knowledge acquired through sense-perception.

62. The Two Purposes which the Senses Subserve.—The manifest purpose of the senses of smell, taste and touch as affected by temperature, is to nourish and protect the body and preserve it in health. The sensation of smell attracts us to the healthful properties or repels from the harmful ones in the food we eat or the air we breathe. The sensation of taste, when pleasant, leads us to select, for the satisfaction of hunger, food that is nutritious and wholesome, and, when unpleasant, to reject food that is hurtful or unfit for the stomach. The sensations of temperature through touch, warn us, by similar conditions, to avoid those extremes of heat and cold which would injure or destroy the bodily organs and to seek such moderation in the temperature of things around us, as to secure our bodily safety. It is for this reason that the hand is not the sole organ of tactual sensation. A network of afferent nerves is actually spread beneath the skin over the entire surface of the body, with the evident design of protecting all its parts from the extremes of heat and cold. Since, then, the purpose of these three senses is solely to guard and preserve our animal organization, we may properly distinguish them as the *animal senses.*

63. Relation of Sense-perception to the Mind. —From the distinction just presented, it is evident that the relation of sense-perception to the mind, is similar to that which sensation holds to the body. Its main office is to furnish the intellect with true nutrition, to supply it with a genuine knowledge of the world with which we are closely connected, and in this way to preserve it from uncertainty and error. Perceptive touch gathers, by contact with innumerable solids, the most definite ideas of their size and shape, ideas that afterwards dwell in the memory with perfect distinctness. Through the perceptive ear the mind collects notions of an unlimited variety of sounds which it recalls at pleasure. Through all our waking hours, the open eye catches and transmits to memory clear mental pictures of the countless forms that surround us in the distance, or near at hand. The notions gained through touch and hearing and sight, are afterwards called up, analyzed, compared and combined with each other; in short, they constitute the materials for the processes of thinking.

Thus it is clear that the main purpose of sense-perception is to transmit to the intellect those primary ideas of outside things on which the imagination, judgment and reasoning are subsequently employed. For this reason we shall call the senses through which sense-perception acts, namely, sight and hearing and touch, *the intellectual senses.*

64. The Pleasures of Sensation compared with those of Sense-perception.—The legitimate use of every organ of the body and every faculty of the mind is accompanied by pleasure, while its abuse is productive of pain. The animal and the intellectual senses are no exception to this general law. The pleasures derived from the two, are, however, widely different in character. The enjoyment afforded by our sensations is in strong contrast with that which attends our sense-perceptions. The gratification which comes from a fragrant odor, is, in no respect, similar to the delight that springs from the contemplation of a beautiful painting.

65. Pleasures derived from Sensation.—But let us examine the facts in the case with more minuteness. A perfume affecting the olfactory nerves, yields a pleasant sensation; a bit of assafœtida, a painful one. The two sensations last only during the presence of the odorous particles that affect the sense of smell. The presence of moderate warmth in the atmosphere begets a sensation of comfort; extreme heat a sensation of distress. But the sensations that bring the most decided pleasure are those of taste. Hunger and thirst beget pressing desires for food and drink, and the acts of eating and drinking, by which these desires are satisfied, give the greatest enjoyment that the animal senses afford. Now a sensation so intense as that of taste, is easily perverted and its enticements lead frequently to habits of excess. Probably the over-indulgence of this single sense especially by intemperance in alcoholic drinks, has caused more wretchedness to the human family than all other perverted bodily sensations together.

But let us note the characteristics of the pleasurable sensations which the animal senses afford.

1. As bodily states they are sensual and carnal.
2. As the mind refers them to the organ in which they originate, they are local.
3. Since the desires, such as hunger and thirst that give them birth, are soon satiated, they are short in duration.
4. They may be called animal gratifications because they are common to man and the brutes.
5. Though their moderate indulgence is needful and right, they are properly classed among our lower pleasures.
6. They are spontaneous and, therefore, need no training but simply discreet guidance and wholesome restraint. Because of a constant tendency to over-indulgence, they require the curb and not the spur.

66. Pleasures derived from Sense-perception.—
The sense-perceptions, however, are attended with pleasures of a far different character. There is, to be sure, only a very moderate gratification in the contacts of perceptive touch with the solids within our reach. But the senses of sight and hearing are never-failing sources of enjoyment during all our waking hours. The ear catches the articulations of language, the accents of love and sympathy, with perpetual delight. Especially do the melodies and harmonies of sound give joys of which we are never weary. The wide variety of notes that wake the tympana into music, add constantly to our happiness. What is more, culture inevitably increases and intensifies these refined enjoyments. To the enthusiast, whose ear is trained to the power of quick discernment of all that is beautiful in music, this enjoyment becomes ecstatic. And yet the ear is never weary, the attention never flags, the perceptive power knows no satiety. The value of a cultivated taste in music lies in the fact that it furnishes intellectual pleasures that are positive, refined and lasting.

Nor are the pleasures arising from sense-perception through the eye less definite in character, while they are far greater in range and variety. Indeed, life owes no small portion of its happiness to the gratification received from the visible world around us. Innumerable objects paint their images every hour upon the retina, and each, unless it be repulsive in form and color, brings its modicum of pleasure. The eye commands distance to a far greater extent than the ear. Within the visual circle, of which it is the centre, it may note at will either one or many objects at a single glance. It may perceive and scrutinize a pebble, a fibre or a leaf which is close at hand ; or contemplate the trees, either singly or in groups, that are farther away; or gather in an entire landscape without effort. And all this it does with wonderful rapidity and without weariness. In this way the sense of sight has the power, far beyond any other sense, to analyze or combine its objects at will, and, in every act of perception, to make its own units. And except in rare instances, every perception it gathers, whether simple or complex, is attended with pleasure.

Now the pleasure so gained will be enhanced just in proportion to the culture of the sense perceiving, and the beauty of the objects perceived. There is actually no limit to the pure enjoyment to be derived by a cultured taste, from the elements of beauty in objects of sight. And this is one of the many reasons why eye-culture lies at the basis of a genuine education. The beauty of objects in the world without is varied and abundant but it is comparatively without meaning and without effect to him who sees in things around him, only what will gratify his grosser appetites. Compare the coarse, narrow pleasures of a savage, who regards the beautiful and sublime in nature with stolid indifference, with those of Ruskin, whose eyes dwell perpetually on all that is beautiful in nature and art. The four great divisions of art, namely, architecture, landscape gardening, painting, and sculpture, are to him treasures whose value never depreciates, sources of elevated delight that are unfailing. How unspeakably sad to go through life with eyes that, from want of training, never perceive the beauty with which this earth is filled.

Noting the special characteristics of the pleasures attending perception through the intellectual senses, we find;

1. That they are not sensual but intellectual.

2. That these pleasures are not localized as in the animal senses, but accompany the ideas gained from without.

3. They are not followed by satiety, and are consequently lasting.

4. They are refined and elevating and not subject to the excesses to which the animal pleasures are liable.

5. They are increased by culture of the intellectual senses and by the improvement of taste in the study of art.

67. The Six Senses; Order of their Growth.—In the order of time the animal senses are called into action and begin their growth before the intellectual senses. Even at birth the wants of the body are immediate and urgent, and the animal senses are at once on the alert to supply them. With the exception of smell, the bodily sensations come into play with spontaneous vigor soon after birth. The sensations of taste and of temperature fulfil the purposes for which they are designed instinctively and without delay, and their growth thereafter keeps pace with the increasing wants of the body they are adapted to serve.

68. Sense-perception later in Activity.—The sense-perceptions, on the other hand, come gradually into conscious activity at a later period. The infant gets his first dim notions of external things only after they have made repeated impressions on his organs of touch, hearing, and sight. The earliest result of these impressions is to beget in the hand, the ear, and the eye, simple sensations only. These sensations awakened by the first contact of light and sound with the optic and the auditory nerves, give place by degrees to sense-perception, which slowly attains distinctness, until the strange new world becomes audible and visible and real.

69. The Order of Growth Summed Up.—Gathering the facts, then, respecting the order of growth in the senses, we sum them up as follows:

1. The wants of the body in infancy are immediate and urgent.

2. Responding to these wants the animal senses are spontaneously active soon after birth.

3. The activity of the intellectual senses produces distinct sense-perceptions at a later period.

4. The first effects of light on the eye and sound on the ear, are pure sensations, which gradually diminishing, give place to sense-perceptions that grow vivid in like ratio, until they furnish distinct notions of external things.

5. The sensations that at first occupy solely the sense of sight and hearing, grow gradually less as perception advances, until in later years we are scarcely conscious of their existence except under an excess of light and sound.

70. The Intellectual Senses—Their Order of Development.—Probably the first conscious intellectual movement of the child is the perception of resistance in touch, and the first dim notions of outer objects seem to arise from their hindrance to the movements of the hand. Nature seems to indicate the antecedence of the perceptions of touch by the constant motion of the infant hands, while the senses of sight and hearing are still comparatively inert. Subsequently, when sight and hearing gather their first notions from the audible and the visible world, they appear, in their earliest intellectual action, to begin and to progress together.

71. The Hand Teaches the Eye.—The qualities of objects as revealed to touch through resistance, are solidity, shape, size, hardness, smoothness, etc., etc.

Now these qualities of matter are made known to us directly through the perceptions of touch only. The eye cognizes directly the modifications of light alone. It perceives *immediately* the colors including what are called light and shade. Light and shade are to the eye only the visible signs of solidity, extension, and shape. The eye does not perceive these qualities but infers them from the varieties of light and shade which their surfaces present. This may be shown in several ways.

1. When the eye cannot distinguish the light and shade of an object because of its distance, its special shape cannot be determined, and it becomes only a dim, visual obstruction.

2. A painter may so represent light and shade on a plane surface that the eye accepts them as indicating an actual solid figure. Thus in many churches the plane walls are frescoed with such skill that they seem to the eye to be adorned with panels and supported by columns. The fitting arrangements of light and shade are there, but not the actual length, breadth, and thickness which they ordinarily indicate. Once visiting a church I observed that its walls were ornamented with beautiful conventional figures that, to the sense of sight, stood out as solid realities. I could not distinguish them from actual shapes except by passing the hand along the wall on which they were painted, and finding it a plane surface.

3. Shut out the light from the room we are in and you have instantly withdrawn from the objects it contains, all colors including the lights and shades. Light and shade, which are the condition of vision, being now absent, the solid bodies present are imperceptible to the eye, and their presence can be determined by the hand alone.

Now arises the question, if the eye cannot directly perceive a solid body, but only the arrangement of the light and shade that lie upon its surfaces, in what manner does the mind learn that these are the signs of the extension in three directions that constitutes a solid? There is but one answer to this question. The hand, which is earlier in action, teaches the eye, by countless instances, that the adjustment of light and shade it perceives, signifies shape. And when the eye is thus taught, every sight-perception is attended by a judgment or inference which is so subtle and rapid as to elude our consciousness. In fact, the perception of light and shade and the inference of shape are so welded together by habit that the intellect accepts them as a single act.

72. The Ear and the Eye Compared.—The sense of hearing, as we have already said, perceives sounds as that of the eye perceives colors and the modifications of light and shade. Sound is the only object perceived by the ear, just as color is the sole object perceived by the eye. And as the mind infers with a subtle and instantaneous judgment solid shapes from light and shade, so does it infer from the various sounds we hear, their external origin, and the causes that produce them.

A bird sings in a neighboring tree, and I instantly infer from the qualities of the sound to which I listen, the direction from which it comes, the place where it is made, and that it is made by a bird. These judgments follow the perception instantaneously, but not with the accuracy or unity which marks our judgments of shape, size and distance from the perception of light and shade. The perception of sound and the judgments of its direction, location and causes, though constant and rapid, are not comprised in one electric flash of thought, like the perception of color and the judgment of shape. The first may be subsequently analyzed into distinct mental acts; the last can scarcely be separated even in thought.

The ear as a perceptive organ is far slower than the eye and the notions of pure sounds we gain through its agency, are fewer in number and far less distinct than those of visible objects gained through sight. It is only when the ear becomes the recipient of significant sounds in language, that it at all compares with the eye as a means of furnishing the mind with ideas, and promoting its growth and discipline. The most effective agent in the early training of the ear, is the vocal organs of the child. As he gradually learns to pronounce from imitation his first simple words, his ear becomes nicer and more discriminating, and readiness in uttering and facility in distinguishing articulate sounds, advance with an equal pace. But this topic will be more fully developed under its proper heading.

73. Specimens of Experience in the Three Kinds of Sense-perception.—Moving about in a dark room, my foot strikes a solid body. I stoop to learn its nature and find, by the mode and direction of the resistance it offers to the motion of the hand, that the upper surface is flat and smooth, and that the surfaces of the sides obstruct motion at different distances but in precisely opposite directions. From these tactual qualities I perceive that it is a leathern trunk. I produce a light and the outlines and varieties of light and shade which the surfaces present are identical with those which the hand and the eye have previously found in conjunction with a similar shape. The hand perceives the trunk from its tactual properties, its manner of resistance; the eye, from its outlines and modifications of color, infers its solidity and shape. The result is a single idea in the mind which contains both tactual and visible qualities.

Again I look out in full daylight and notice an object, say two hundred yards away. The outlines and changes of light from its surfaces, enable me to recognize it as a horse. Now how do I determine its distance? Among the various means of judgment there are three prominent combinations.

First; the comparative size of the image it makes on the retina.

Second; the comparative dimness or distinctness of its outlines and colors.

Third; the things that intervene between the eye and its object.

Remove or change any one of these conditions that ordinarily attend the acts of vision, and the judgment is unreliable and incorrect. Look, for example, at a flock of birds flying above us and having, therefore, no intervening objects; and your judgment of their distance is therefore illusory.

Again; traveling twenty-five years ago on the western prairie, one morning we came in sight of Chimney Rock and supposed from its appearance that we could pass it in two hours' travel. We actually traveled all day and camped over night at its base. Our judgment of distance was in this case led astray in all three particulars.

1. The level plain furnished no intervening objects.
2. We had no previous information of its actual size and height.
3. The remarkable clearness of the atmosphere in that region revealed its outlines and brought out its irregularities of surface with wonderful distinctness, thirty miles away.

Another instance shows that a mistake as to the dis-stance of the object, may vitiate our estimate of its actual size. Professor Williams, formerly of Michigan University, was walking one morning along a path parallel to two close board fences, the first of which was some forty yards from the path and the second fifty yards from the first. The tops of the fences were on a level with his eye and when a fowl which had been invisible between the two, flew onto the nearest one, he instantly referred it to the more distant. The professor said that the size of the fowl was astounding and that it looked like some monstrous bird from another planet.

Let us now scrutinize the judgments that attend the perceptions of sound. I hear the discharge of a pistol. From the quality of the sound coupled with my former experience, I know it to be a pistol-shot. Now what elements have I by which to infer its direction and distance?

1. The greater effect it makes on one ear rather than the other, suggests the point of compass from whence it comes.

2. Knowing that the sound is a pistol shot, its comparative loudness suggests to me its distance and locality. These criteria, it must be confessed, are variable and uncertain, and consequently such judgments are often false.

In many instances the eye gives important help to its sister sense in determining the course and locality of sounds, especially the more distant ones. For example, we hear the rumbling of the cars and often cannot tell whether an east or west going train is approaching, until a glimpse of a distant smoke or headlight decides the question. We judge of the remoteness of thunder by the length of the interval between the peal and the flash.

QUESTIONS ON CHAPTER III.

What two purposes do the senses serve? Give examples of the senses of taste and sight. Name and describe each of the organs of the senses. What is the difference between sense-perception and sensation? Define and illustrate sense-perception. Define and illustrate sensation. The uniform connection of knowledge and feeling. Describe the contrast of sensation and perception in consciousness. Illustrate the senses of smell, taste, and touch; and show that every sensation is either pleasant or painful. Show by example that touch reveals the qualities of extension in matter, by contact. Show by example how the ear gains knowledge of an external sound. No consciousness of sensation in the operations of sight and hearing. What three senses yield sensations, and what three are instruments of perception? Compare sensations and perceptions in memory, and say whether they are distinct. Which do we remember most distinctly, the acts of smelling and tasting or the sensations they produced? Recall the products of sight, hearing, and touch as a perceptive sense; and compare their distinctness with those of sensations. How does your memory of a familiar face compare in distinctness with the memory of the odor of a rose? Paucity of terms in the language which designate sensations. Give examples. Scantiness of terms denoting the sensations of touch. Double meaning of such terms. Number and distinctness of terms that designate ideas gained by sight, hearing, or perceptive touch. Give examples of their abundance. The manifest purpose of the senses of smell, taste, and touch. Why are the three senses that produce sensations called animal senses? What is the main office of sense-perception? Give the character of the ideas which each of the three senses of sight, hearing, and touch gathers for the mind. By what name should sight, hearing, and perceptive touch, then, be distinguished from the animal senses, viz., sensitive touch, taste, and smell?

On the Intellect. 55

Difference between the pleasures derived from the animal and from the intellectual senses. Illustrate the pleasure derived from a sensation naturally produced. Compare the pleasant and the painful sensations, and show how each is produced. Show how sensation is easily perverted. Describe the pleasures derived from the senses of sight, hearing, and perceptive touch. Give the six characteristics of the pleasurable sensations which the animal senses afford. Range and variety of the pleasures derived from sense-perception through the eye. Power of analysis and combination possessed by the sense of sight. Pleasures gained from hearing and sight in proportion to their culture. Pleasures of the savage compared with those of the artist. Give the five characteristics of the pleasures which attend sense-perception. What senses are first in the order of their growth? The intellectual senses later in their action. The intellectual senses begin in sensations. Give the five facts that present the order of growth in the senses. Perception of resistance in touch the first intellectual movement of the child. Show how the hand teaches the eye the qualities of solidity, shape, etc. The eye sees color directly, and infers the qualities of solidity through light and shade. Give the three ways in which this is shown. Sight-perception always attended by subtle judgments of size and figure. Sound the sole object perceived by the ear. The causes of sound an inference from its effect on the ear. Comparative slowness of the sense of hearing. Effective agents in the early training of the ear. Example of the action of touch in a dark room. How do the hand and the eye act in concert? By what three combinations does the eye judge of the distance of objects? Why did my judgment fail in estimating the distance of Chimney Rock? Prof. Williams' mistake in judging the size of a fowl. By what two elements can I judge of the direction and distance of a pistol-shot? Help of the eye in discriminating the distance of sound.

Chapter XV.

INTERNAL PERCEPTION—SELF-SCRUTINY.

74. Internal Perception Defined.—Internal perception is the second faculty by which the mind acquires new knowledge. Sense-perception, as we have seen, gathers knowledge of the facts which are without. Internal perception gains knowledge of the facts that are within us. The one perceives the qualities of matter; the other the present thoughts, feelings and decisions of the mind. In the one the eye of the body is directed to visible objects; in the other, the eye of the mind is turned upon its own acts and states. I see the flying clouds and falling rain, and hear the distant thunder. These, as we have learned, are acts of sense-perception. I recollect a similar storm that occurred a week ago, that is, I perceive that I am thinking of a particular past event retained in memory. This is an act of internal perception.

It is, of course, unnecessary to say that internal perception acts through no organs which are analogous to those of sense-perception. It is the simple power which the mind has of scrutinizing more or less intently its own present acts and states. Thus I recollect that when the steamer in which I crossed the ocean, was lying in the harbor, a man fell overboard, and the mate, leaping into the water, rescued him from drowning. These vivid pictures of a past event I recognize by an act of internal perception, and, along with these pictures, arises also the same feeling of admiration which the heroic act I witnessed, excited at the time. The act by which I perceive this feeling, not only as a past experience but as a present mental state, is also an act of internal perception.

Heed should be given that we do not assign this faculty too narrow a province. It comprises the whole field of our mental experiences. It takes note of every conscious movement in memory, judgment, imagination, and reasoning. It marks and identifies every feeling in the wide range of sensations, passions and emotions of which we are capable. It distinguishes each one of the innumerable efforts of will that are put forth daily. All these individually, as they flash through the mind, together with their order, connection intensity and effect, are the object of internal perception.

This power by which the mind knows its own acts and states, is inherent in every mental movement, is inseparable from that movement and is essential to its very existence. For without knowing that we think or feel or will, we cannot think or feel or will at all. Internal perception is, when strictly defined, *simply consciousness directed by the will* to any operation or state of mind. The distinctness with which the mind perceives its intellectual acts, will of course depend upon the degree of disciplined power which it has previously attained. The ability to scrutinize through consciousness the character of our own mental processes, to note accurately each step in the complex operations of thought, is a rare and valuable attainment. It is well to observe that the thoughts, the feelings and the efforts of will that constitute the objects of internal perception, have no relation to space. They occur in time but, since thinking and feeling and willing have none of the properties of extension that belong to external bodies, they cannot be located in space.

75. Attention, Observation, Reflection.—Attention is the voluntary effort which any intellectual faculty puts forth to gain a knowledge of its appropriate object. It is the energetic application of an intellectual power to its peculiar purpose, namely, the attainment of complete knowledge in the line of its special activity. The etymology of the word, *attention* —*attendere*—to stretch towards—suggests its meaning as applied to an act of mind. Attention is a strenuous mental act—an act impelled by the will, in which the intellect concentrates exclusively upon its object. In attention, the mind listens intently through the ear, scrutinizes earnestly a visible object through the eye, fastens with exclusive vigor on the memory of a past event, etc., etc.

Attention is indispensable to study, investigation, research, in short, to every form of successful mental labor. Every teacher knows that those pupils progress most rapidly who are capable of giving earnest attention in solving the difficulties which a lesson presents. The first requisite in the solution of a complicated problem, is to fasten the thought upon it with unyielding tenacity. The chief characteristic in the intellect of a savage, is that he is incapable of prolonged attention. The attention of a young child is weak and fitful and soon wearies; the progress of education is marked by a constantly increasing power to apply the mind steadily to its various objects. Other things being equal, the highest mental power attainable to man, is evinced by the acquired habit of bringing and holding the intellect to a focus upon the subject under scrutiny. Sir Isaac Newton regarded himself as superior to other men only in the power of prolonged and intense application. Socrates could stand motionless for hours, with closed eyes, oblivious of surrounding objects, pondering on the problems of human destiny.

76. Observation.—Observation is a series of connected acts of attention, exerted through the intellectual senses, upon external things, their qualities, relations and changes. Hence in science, observation consists in repeated and systematic efforts of attention directed to the discovery of some property or law of material objects. Scientific observation notes methodically those unvarying qualities of matter that serve as means of scientific classification. It follows that the material sciences are largely the offspring of observation. The stereoscope, telescope and microscope are its instruments.

77. Reflection.—Reflection, on the other hand, is the continued and strenuous attention of the mind to its own acts and states. It is the voluntary focusing of consciousness upon the mental operations, in other words, the energetic application of internal perception to its objects. The original meaning of the word, *reflection*, is to turn back. In every act of reflection, the mind turns back against the usual current of thought, to scrutinize its own phenomena.

As I sit by my table, there springs up in memory, unsought, a picture of the great military review I witnessed at Homberg. I turn the eye of consciousness upon it and consider it attentively, in order to determine the occasion of its presence in my mind at this moment. Out of the depths of memory, it rose spontaneously to the surface of consciousness. Scrutinizing the mental picture, I find that its presence is due to the music of a military band I had just heard in the distance.

These are the simple acts of reflection wherein consciousness impelled by the will, scrutinizes, through internal perception, its own thoughts and feelings. When trained to high power of sustained self-scrutiny and acute discrimination, it is the province of reflection to detect the facts of mind that are scientific in character. In this regard it corresponds to scientific observation in the outer world. The products of philosophic reflection are such sciences as the pure mathematics, rhetoric, logic, and psychology. It is evident that disciplined reflection, the ability to examine exhaustively one's own mental processes, ranks high in the scale of intellectual attainments. It is the rarest and most difficult of all our intellectual achievements, and can be acquired only by patient and repeated efforts of the will extending through years of practice.

The list of names which history has presented of those who possess the power of self-scrutiny in a remarkable degree, is comaratively scanty, and the men of our country who have attained to eminence in this direction, can be counted on the fingers. All persons, however, of ordinary intellectual capacity, possess this power to a limited extent. Children and savages, are wholly unable to arrest and inspect the current of thought which naturally sets outward. To the uncultivated and the ignorant, whose desultory attention is wholly occupied with the objects presented to the senses, the effort of reflection is wearisome and painful.

78. Order of its Development.—It is evident that, in the order of intellectual growth, the power of reflection is the last to be developed. The acute intellect and the disciplined will which absorbing self-study requires, can be attained only after the other intellectual faculties have reached the ripeness of their strength. It will be seen that observation and reflection require trained acuteness: the one of the intellectual senses, the other of internal perception. A will that is facile, powerful and capable of long-continued efforts, is indispensable to both.

QUESTIONS ON CHAPTER IV.

Internal perception—what is it? Give the distinction between internal perception and sense-perception. What are the objects of internal perception? Its field comprises all our mental experiences. Is there any distinction between internal perception and voluntary consciousness? Internal perception a rare and valuable attainment. Define attention. Attention indispensable to effective intellectual action. Attention feeble in children and savages. Attention one of the highest powers attainable. What is observation? What is the purpose of scientific observation? Give the essential characteristics of reflection. Examples. Distinction between reflection and observation. Trained power of reflection a rare attainment. The ignorant man lacking in the power of reflection. The power of reflection latest in its development.

Chapter V.

MEMORY—RETAINING CONCEPTS.

79. Order of its Activity.—The faculty of memory stands second in the order of intellectual growth. Memory is awakened in infancy by the knowledge gained through sense-perception. As the intellectual senses are incited to their first conscious activity by the presence of sounds and visible things, so the first movements of memory are called forth by the objects which these senses have repeatedly presented to it. The earliest activity of both sense-perception and memory is spontaneous. The first is stimulated to action by outer objects; the second by the concepts gained from these. The order is therefore invariable. First to acquire knowledge and then to retain it, is a succession which, in the nature of things, cannot be inverted. We cannot retain what we have not first acquired and the act of retention is necessary to complete the act of acquirement.

80. Varieties of Memory.—But while its order of development is fixed, the varieties of memory found among men of like mental calibre in other respects, is remarkable. No intellectual faculty shows, in different individuals, such a striking diversity of attainment in special lines. Many men of great intellectual power have been deficient in verbal or local memory. Joseph Scaliger, the most learned man of his day, found difficulty in recollecting proper names, while Cyrus, whose scholarship was inferior to Scaliger's, could repeat the name of every soldier in his army. Many persons have an excellent memory of faces while utterly lacking the power to recall the particular names that belong to them. I have known men whose capacity to retain scientific facts was exact and sure, while their memory of names and dates was fickle and uncertain. Ben Jonson declared that he could repeat all that he had ever written and whole books that he had read. Most authors that have written much, can scarcely repeat with exactness what they have written, beyond a few sentences. Occasionally a man is found who, with ordinary ability in other respects, shows a remarkable memory in a special direction. Dugald Stewart mentions an individual of his acquaintance who, though completely ignorant of Latin, could repeat thirty or forty lines of Virgil after having heard them once read. He was, however, noted in every family in which he had been employed, for habits of forgetfulness, and could scarcely deliver an ordinary message without committing some blunder.

Numerous instances may be found of men deficient in other striking mental traits, who show a marvellous quickness in memorizing names and even unmeaning articulate sounds. Others seize instinctively and retain permanently the meaning and thought of the author, forgetting his words. Another class astonish us with the lightning rapidity with which they commit columns of figures and arithmetical calculations, which they retain and recall without seeming effort. Others still, like Jay Gould, show marvellous facility in mastering and remembering innumerable details of business. And others, like Hamilton, retain and reproduce with clearness the whole wide variety of philosophical opinions given to the world from Pythagoras down.

81. Value of Memory.—The varieties of memory

mentioned above are widely different in their relative importance. Mere feats of mnemonic jugglery are of little account while the memory that furnishes promptly the needed material for judgment in the various pursuits of life, is of untold value. Great and striking as are the individual peculiarities that distinguish men otherwise on the same intellectual level, the fact remains that memory is largely the creature of education and habit. Beyond all the other faculties, it will respond to the demands that are persistently made upon it.

82. Memory includes Three Acts.—An accurate analysis of memory reveals that it is a complex faculty embracing three closely related yet distinct acts, namely, the act of acquiring, the act of retaining and the act of recalling. Writers on this subject have not generally regarded the process of acquiring knowledge as an act of memory. But since it can recall only the knowledge which it has first received, no valid reason can be given why memory should not include acquisition and recollection as correlative acts, the one being the indispensable antecedent to the other. Indeed ordinary language recognizes the act of acquiring as an operation of memory, by such expressions as "committing to memory," "storing the memory." If it be said that the act of acquiring is the uniform result of the effort of another faculty, the reply is that this is true also of the act of recalling.

83. Order of Action.—Evidently the processes of acquiring, retaining and recalling, follow invariably the order in which I have named them. We cannot retain without having acquired, nor can we recall without having retained. It is clear also that each process depends for its completeness, upon that of its antecedent. Thus we recall vividly only what we have retained perfectly. We retain perfectly only what we have acquired thoroughly. We acquire thoroughly only what we perceive clearly and exhaustively. Every complete act of memory, then, must be preceded by the effective action of the intellectual faculty that supplies its material. Thus, visiting the gallery of the Louvre, I stand before Murillo's great painting of the Conception. I acquire a complete idea of the figures represented in proportion as I contemplate them earnestly; I retain them perfectly if complete when acquired; I recall or recollect them distinctly in proportion as they have been retained perfectly, acquired completely, observed exhaustively. The vividness of recollection depends on the efficient action of the preceding faculties, namely, retention, acquisition, sense-perception. Completeness of retention depends on thoroughness of acquisition and sense-perception; while acquisition depends for its completeness on the earnest effective effort of sense-perception which is its immediate source.

Hence it is clear that while every distinctive act of mind should itself be strenuous and energetic in order to be effective, the initial act by which knowledge is gained, is by far the most important. Notions gained through the intellectual senses, which contain in completeness all the properties of the objects they represent, afford the genuine material for an active memory and a sound judgment; while vague half-formed notions, which are the product of desultory attention, tend constantly to vitiate memory and weaken judgment. While each intellectual faculty must have frequent and systematic exercise as a means of growth and vigor, nothing can compensate for a deficiency in accurate knowledge. It follows that the systematic training of the intellectual senses in early life, is vitally essential to the subsequent discipline of the memory and the faculties of later growth.

84. Attention Necessary to Acquisition.—The first requisite to thorough acquisition is strenuous attention. If I examine by reflection the contents of my own memory, I find that those concepts which I have gained by absorbing and repeated acts of attention, are most vivid, distinct and easily recalled. The only picture of a collection in Liverpool which I clearly recollect, is Turner's great painting of the burning Parliament House, which I studied with exclusive earnestness. I remember, even to minutest particulars, the group of bronze figures that adorns the Schiller Platz in Frankfort-on-the-Main, because I examined them carefully and frequently. I can now repeat with ease and accuracy Hamlet's soliloquy, because years ago I made an earnest effort to commit it to memory. Of the books that I have read, I can reproduce the contents of those only to which I have given the greatest heed. Of the difficult mathematical processes, I find that I recall most completely throughout, those which I thoroughly mastered by fastening upon them with the grip of unyielding attention. From infancy to old age the law holds good that knowledge acquired, depends for its accuracy and completeness, on the degree of attention expended in gaining it. Those men whose knowledge is remarkable both for extent and minuteness, are also remarkable for a habit of concentrated attention on the subject in hand. It is safe then to repeat that trained attention is a prerequisite to a precise and ready memory and therefore to intellectual culture.

85. Selection in Acquiring.—It is also important, if we would store the memory with permanent treasures, that we make judicious selections of the objects of knowledge. No earnestness of purpose or strenuous efforts will avail if we expend them upon trifles. Many a mind is enfeebled beyond recovery, by giving exclusive heed to things and events a knowledge of which is valueless as a means of intellectual culture.

In ordinary life objects and incidents are presented to the intellectual senses in the promiscuous manner that belongs to mere local connections. He who observes these in their local relations, without investigating the permanent qualities by which they are arranged in different classes, uses his memory as a miscellaneous scrap basket, instead of a well-ordered counting-room where every paper is numbered and filed. Walk out any morning in summer and you will see perhaps, among the many objects that strike the eye, a few butterflies and several birds of different plumage. Give to these a passing notice as mere animate creatures of different kinds, and the desultory knowledge you have gained, will either wholly elude the memory, or lie in it as useless lumber. But if you are able to note in case of each the unvarying peculiarities of color, form and organs that characterize its species and separate it from others, in other words, to classify each bird and butterfly instead of giving it a random glance, you will add ideas to the memory that are worth the gathering.

That man is fortunate who, be his avocation what it may, is proficient in at least one science that systematizes the common objects which his senses hourly encounter. For the acquisitions that are limited to the trifling objects and petty incidents which are connected only by the accidental relations of time and place, are sure to result in a fickle memory and a feeble intellect.

86. Special Inclination.—Special inclination or

bent of mind has much to do with the facility with which each person acquires new knowledge in his own line of observation. The shepherd knows every individual sheep in his flock and distinguishes at once the slight peculiarities that separate each one from all the rest. The horse jockey decides at a glance the qualities of a horse, its capacity for speed and draught. This facility of acquisition in special lines, which belongs to the expert, is due to a power of attention whose discipline has been confined exclusively to a single class of objects and their relations, and to the exceptional pleasure which the observation of these objects affords. Objects of natural interest stimulate the attention and render acquisition rapid and accurate. We learn easily and retain surely what pleases and interests us.

87. Beauty helps Acquisition.—Now the quality in things observed that is most attractive, is that of beauty. Whether the element of beauty be addressed to the eye, the ear or the judgment, attention fastens upon it with instinctive energy. How readily any child learns the series of melodies which constitute a simple tune. How readily the successive notes of a familiar air recur to memory. In place of the harmonies or, in other words, the beauty of sounds, try the ear with a similar recurrence of unmusical noises. It is doubtful if you could learn and retain them at all in the order presented. The ear shrinks from such a series with natural aversion. For a like reason, namely, the music of measure, one commits and recalls poetry with less difficulty than prose. Every one knows how gladly we distinguish and dwell upon the objects of beauty in nature or art, that present themselves to the eye. A landscape garden impresses itself upon the memory more distinctly than a like extent of uncultivated, monotonous landscape.

Objects of natural beauty and the products of art constantly appeal to our intellectual senses and render the effort of attention by which the mind masters their details, earnest and absorbing. The artistic treasures collected in the great cities of Europe, attract annually thousands of those who regard no sacrifice too great for the profit of beholding whatever is exquisite in form and color. The quality of beauty in all its aspects, is one of the prominent means of educating sense-perception and storing the memory with valuable knowledge.

88. Novelty helps Acquisition.—Another circumstance that facilitates the act of acquiring new facts for the memory, is novelty. The human mind has a natural longing for whatever is new and strange. This eager desire for novelty, which is strongest in childhood, is called *curiosity*, and its gratification excites the emotion of surprise which gives to attention a powerful impulse. The prevalence of an eagerness to see what is unusual and rare, is indicated by the crowds that throng the menagerie and the circus. It is one of nature's provisions for extending our range of knowledge, and the ideas gathered under its influence, naturally cling to the memory.

89. Strong Feelings help Acquisition.—Whatever forcibly excites emotion or passion, stimulates attention in a high degree and renders clear and full perception inevitable. Knowledge gained under the impulse of intense feeling, is of all our mental acquisitions, the most vivid and permanent. We never forget the events that cause us great joy or sorrow; the circumstances of extreme personal peril, the incidents that arouse to a high pitch, fear or anger, hope, pity or dismay. We could not efface them from memory if we would. They become in fact the central notions, around which our ordinary experiences are wont to cluster.

90. Retention—Possible Knowledge.—Though we have named retention as one of the three acts which memory embraces, it is not strictly an act. Retention is passive rather than active, is a capacity instead of a faculty. Its province is simply to hold fast to the concepts which acquisition supplies. But the knowledge retained is possible knowledge and not actual knowledge. That knowledge is called possible which the mind holds in its possession but not under its eye at the moment. Actual knowledge is knowledge recalled and positively contemplated by the mind at any given instant. In solving an arithmetical problem, each step, as you consider it, is actual knowledge, while your entire acquisitions in arithmetic, are preserved in memory as possible knowledge. Actual knowledge is that of which we are conscious just now; possible knowledge is that which we hold in unconsciousness.

Retention then is the depository of possible knowledge. It is the unconscious interval between two acts, namely, the act of acquiring and the act of recalling, and in crossing this interval, be it an instant or a decade, percepts are changed into concepts. Thus a concept, whether of a concrete thing or an abstract quality, is a percept resuscitated. Such concepts, modified by analysis or classification, are reconsigned to retention, the realm of possible knowledge, to be restored or summoned again when the necessities of thought require their presence.

91. Possible Knowledge Naturally Transient.—It is a law of our nature that possible knowledge consigned to memory in its retentive capacity, tends constantly to fade. This process of effacement by time continues, unless arrested, until knowledge passes wholly beyond the limits of possible recall. Especially is this true of all concepts that were feeble or indistinct when acquired. Of the countless acts of perception which go to make up the mental experience of a single ordinary day, not one in a thousand abides in memory. In general, it may safely be said that ordinary objects and incidents associated by the accidents of time and perceived but once, add nothing to the prominent accessions of the memory. How many of the faces that we see in a crowd, do we reproduce subsequently? How many rambling remarks that we hear in a desultory conversation, can we ever repeat? The fact is, memory selects for permanent possession only those experiences which rise above the level of the dull uniformity, from which the highest human life is by no means free. This sifting process, rigidly conducted according to principles explained under acquisition, goes on unceasingly. It is a wise provision for preserving those notions exclusively that are worth remembering. Otherwise memory would be a mere *omnium gatherum* filled to repletion with insignificant trifles that would perpetually clog the judgment.

92. **Possible Knowledge—How Preserved.**—The pictures drawn in our minds says Locke, are laid in fading colors and, if not sometimes refreshed, vanish and disappear.

By what means can we retard the fading pictures and preserve their freshness? In what manner can the mind, by its own action, counteract the gradual effacement of time? The plain answer is that the concepts of memory can be retained in their original distinctness only by frequent renewal. The mind preserves its memories from the dimness which time would otherwise produce, by repeated acts of scrutiny. Possible knowledge may be rescued from decay by changing it, at short intervals, into actual knowledge. In other words, the concepts which the mind retains in the darkness of unconsciousness, must be frequently recalled to the light of consciousness, in order to revive their distinctness. For the intellect restores to its first vividness every concept which it earnestly contemplates.

93. Two Methods of Renewal.—There are two methods of renewal by which our possible knowledge may be preserved from fading and from final obliteration.

1. By repeating the acts of perception by which it was originally acquired.

2. By repeated acts of recollection.

As to the first, it is clear that the knowledge which is gained and made complete by reiterated acts of attention, becomes, in the highest degree, familiar and permanent. I can never forget the objects with which I have come into frequent contact for years. The faces and forms, even the voices of those by whom we were surrounded in childhood, are held tenaciously in memory through life. In like manner the elementary knowledge acquired in the primary school, being frequently reviewed in subsequent progress, becomes finally well-nigh unfading. No lapse of time or isolation from books could, for instance, efface from memory the letters of the alphabet. So it is with all knowledge. When once acquired, it is held in memory with a tenacity and distinctness which, other things being equal, are in proportion to the number and frequency of repeated efforts put forth in acquiring it. This is nature's method of training the memory in childhood. In early life the intellectual senses are the sole avenues of acquirement. At this period the currents of mental activity set outward and render innumerable repetitions of effort inevitable.

94. Value of Reviews.—To the thoughtful teacher, this natural method of strengthening the capacity of retention in the young, suggests the value of frequent reviews. Of the more formal knowledge acquired in the processes of education, that is most vivid and available which has been perfected by repeatedly reviewing it. But the most effective means of preserving the concepts of possible knowledge in their original clearness of outlines, are the acts of frequent recollection. Knowledge which is often recalled from retention and scrutinized, maintains its vivacity in a high degree, and becomes thereby the material on which thought is daily employed. It is true in most cases, especially in studying the various branches of knowledge, that successive acts of distinct recalling demand a greater mental effort than a repetition of the act of acquiring. Studying human physiology, for example, and learning the different nervous centres, their names and uses, one might, after an interval, more easily repeat the effort of acquisition than that of recollection without the aid of a book. But this latter act implies, of course, great thoroughness of acquisition.

95. The Retentive Capacity the Storehouse of all Knowledge.—In explaining the acquisitive acts through which the retentive capacity receives its stores, the illustrations employed might seem to imply that the only source of the knowledge retained in memory, is the intellectual senses. But while sense-perception is the main avenue of supplies to the memory with children and with men whose attention is absorbed in external matters, it is nevertheless true that the internal as well as the external worlds, contribute to the sum total of knowledge retained. Through internal perception memory gathers a portion of its concepts and judgment, imagination, reason, and the entire sensibilities commit their products to its keeping.

96. Recollection the Act of Recalling.—We now reach the final act of memory, that of recalling the concepts held by the retentive power in the great realm of possible knowledge. After what we have said, it is hardly necessary to remind the student that in general the readiness and ease with which concepts are recalled depend on the completeness of their previous acquisition and retention. But it will appear, as we proceed, that there are important special conditions which greatly affect the facility and certainty of recollection. These consist in the manner in which concepts held in retention, are connected with each other. In other words, promptness and certainty in recalling any given concept, depend on its mode of association with other concepts.

97. The Laws of Association.—The laws of association by which ideas are held in memory and restored to consciousness, are as follows:

One idea suggests another when both are associated together by any of the following relations, namely:

1. Time. 2. Place. 3. Whole and parts. 4. Sign and thing signified. 5. Cause and effect. 6. Means and ends. 7. Resemblance. 8. Contrast. 9. Accidental relations of whatever kind.

98. Associations of Time.—When two or more events under observation occur either simultaneously or in immediate succession, they are thereafter associated, so that when one is recalled, the other instantly comes to mind also. This association does not require that events should stand in any other relation to each other than that of time. If they are connected together as cause and effect, means and ends, etc., etc.

such relations are independent of the order of time in which they occur. The associations of time as a means of recalling events, require simply that they should be perceived as occurring together or in close succession.

The following may serve as one of the simple examples of this mode of association. Returning to my lodgings late at night, I heard suddenly the ringing of bells and the cry of fire. Soon thereafter came the clatter of hoofs upon the pavement and the fire engine passed at full speed. Following closely half a dozen blocks, I caught sight of the flames bursting out from a neighboring roof. Here we have several incidents presented to the senses of the observer as occurring one after another. Whenever subsequent experience recalls any one of these to conscious memory, the recollection of all the others will immediately follow in the order of their occurrence.

A little reflection will enable us to see clearly that the associations of time are the means by which we retain and recall the ordinary experiences of life. For this reason, the different parts of a narrative are connected by the order of time in which they occur. Observe this order in the following passage.

"On the following morning very early, as the schoolmaster stood at his door, inhaling the bright wholesome air and beholding the shadows of the rising sun and the flashing dew-drops on the red vine-leaves, he heard the sound of wheels and saw Mr. Pendexter and his wife drive down the village street in their old-fashioned chaise known by all the boys in town as the ark."

99. Associations of Place.—No one who gives the least attention to his own mental experiences, can fail to notice that a large share of the concepts he recalls to consciousness, are images of objects that are grouped together in space. One sees, further, with a very little reflection, that his most vivid mental pictures held in memory, are the images of places and the objects they contain. If a prominent object in a familiar locality be brought to mind, all the other objects which cluster around it, at once reappear. The flash of recollection that restores the whole group to consciousness, whenever a single member of it is the object of thought, is spontaneous and irresistible.

Call to mind the house in which you passed the early years of childhood, and the mental picture you recall is the house and its surroundings. You cannot, even by the most strenuous effort of will, exclude these surrounding features of the familiar landscape and contemplate the house as a solitary image. Fix the mind's eye upon any single article of furniture in a well-known room from which you are absent, and the entire apartment with its complete equipments is instantly before the mental vision.

Or visit Westminster Abbey and scan the many artistic figures and inscriptions that adorn the monuments of the illustrious dead, who rest beneath them. Try to recollect subsequently a single inscription, figure, or tombstone, without restoring, with more or less distinctness, the adjacent objects of like character and your effort will be in vain. Whenever any element of locality which the memory retains, is pictured in consciousness, the mind spontaneously completes the whole picture.

It will be noticed that these mental pictures are all composed of what may be termed sight concepts or concepts derived from the sense of sight. These sight concepts, which are far more numerous than those gained through the other intellectual senses, are held in memory largely under the associations of place. As will be seen hereafter, they are also classified mentally under the relations of resemblance, cause and effect, etc., etc.

100. Time and Place in Conjunction.—Events are contained in time. Material objects are located in space. Now since, to the observer, time and space are always conjoined and objects and events are presented to the senses together, the associations of time and place are always united in the mind and will consequently be recalled simultaneously. Any descriptive passage will show how closely connected in memory, are these two modes of association. The following extract taken at random from "White Wings," by William Black, is an example.

"Storm indeed! when we scramble upon deck again, we find that it is only a brisk sailing breeze we have, and the White Dove is bowling merrily along, flinging high the white spray from the bows. And then we begin to see that, despite these dreamy mists around us, there is really a fine clear summer day shining far above this twopenny-ha'penny tempest."

The opening lines of "The Lady of the Lake" furnish a happy instance of the manner in which time and place are blended in objective poetry.

> "The stag at eve had drunk his fill,
> Where danced the moon on Monan's rill,
> And deep his midnight lair had made
> In lone Glenartney's hazel shade:
> The deep-mouthed bloodhound's heavy bay
> Resounded up the rocky way,
> And faint from farther distance borne,
> Were heard the clanging hoof and horn."

101. Association of Whole and Parts.—Every whole thing which has been an object of perception, is associated in memory with its parts. Whenever one is recalled, it inevitably suggests the other. Our notion of a carriage, for example, comprises its wheels, body, springs, seat, top, etc. Now if at any time a notion of the whole is suggested, the parts of which it is composed, are brought to mind at the same time. On the contrary, if any single part as wheel or top, is brought to mind, it at once recalls the whole. My concept of the human body contains as parts, notions of head, trunk, limbs, hands, feet, etc., etc. Now a recollection of the whole contains the subordinate recollection of the parts, or the recalled notion of any part brings to mind the entire body.

102. Sign and Thing Signified.—It is the association of sign and thing signified, that renders a language of articulate sounds, possible. The ability to make an arbitrary articulate sound the sign of an idea, a thing or class of things, so that they suggest each other with inconceivable quickness, is one of the most wonderful powers of the human mind. The faculty of speech which is based on this subtle association, distinguishes man from the brutes and enables him, as will be seen farther on, to arrange his concepts into classes and to perform the operations of judgment and reasoning.

The facility with which the associations of words with things is employed in acquiring a spoken language, varies greatly in different periods of life. Children catch and associate new words and their objects without conscious effort, and learn to talk with rapidity. Later on the ear becomes less facile while the power of utterance even increases. A speaker, at an ordinary rate of delivery, pronounces about one hundred and twenty-five words a minute. In the juvenile mind, the memory of words is more susceptible and the power of acquisition more active than in the adults. In the latter, the act of acquiring is less spontaneous, while the act of recalling is ready and rapid. In advanced life, the power of acquiring words becomes sluggish, but the power to recall them is less impaired. It is worthy of notice, moreover, that most minds pass from the sign to the thing signified more easily than from the thing signified to its sign. In other words, the name suggests its concept more readily than the concept its name. This is undoubtedly due to the fact that when the word suggests the object it denotes, we gain such word through perception either from the tongue of another or from the printed page, whereas when the concept suggests the word, it is itself usually suggested by a kindred concept.

In reading or listening to language, words recall ideas; in speaking or writing ourselves, ideas recall words. This is a reason why students have, in conversation, so little command of a foreign language which they learn solely from books. Another reason would seem to be, that the language acquired from the printed page, supplies the memory with the forms of words, while the language of conversation is composed of their sounds. It is evident from these facts, that, in studying a living language, whether foreign or vernacular, the eye, the ear and the tongue should have full practice, so that, when mastered, it may answer equally the needs of both acquisition and expression.

103. Cause and Effect.—Causes and their effects are associated in memory and mutually recall each other. Thus heavy clouds suggest rain; a frost in early autumn, injury to the crops; falling temperature, ice; the presence of a fatal epidemic, death. On the other hand, the thought of snow recalls a thought of winter; the charred remains of a dwelling, the thought of fire; a pale face and wasted form, the thought of disease; the various products of the farm, the thought of the soil that produced them. A cause and its effect, being united by a close relation, are always connected in experience, and, consequently, are uniformly associated in memory. Since any given cause is repeatedly observed as the antecedent of an invariable effect, its association with that effect in memory is a permanent one, though less vivid and facile than that of sign and thing signified. Because of the fact that this relation of cause and effect is immutable, it has, as will be seen hereafter, been made the basis of classification in certain branches of science.

104. Resemblance.—The most extensive and available of all our means of recalling ideas in memory, is that of their resemblance to each other. That one object, whether perceived or conceived, suggests another which is similar to it, is one of the most familiar facts in psychology. The most trivial and the most important incidents, the innumerable host of objects, great or small, that daily fall under our observation, excite the recollection of kindred acts and objects of previous experience. A portrait suggests its original; a soldier, his fellows; an expression of face in a stranger, a similar one in a friend; a successful general, a number of like characters in history. In short, the relation of resemblance connects all the objects of human knowledge whether material or mental. For this reason it is manifestly impossible to dwell upon a single concept of memory without suggesting others that are similar.

Indeed nearly all the words of a language are the names of greater or smaller groups of thoughts, things, or acts that are connected together by the relation of resemblance. Thus the word, *man*, denotes a vast number of similar creatures that resemble each other in their important characteristics. *To walk*, is the name of a multitude of acts agreeing in certain particulars. So boy, horse, tree, house, forest, to sing, speak, strike, think, feel, eat, are instances of words that denote great groups of things or acts that are arranged under the relation of similarity. Pronounce the word and it tends to recall the group of objects its signification covers. Or the thought of any member of the group excites instantly the thought of the names of other members. Resemblance in constant qualities is the basis also of the many classifications in science. The plants in botany, and animals in zoology are classified mainly on similarity of origin and organs.

105. Contrast.—A great number of contrasting objects and acts which are closely connected in our mental experience, suggest each other in recollection. Life and death, heat and cold, far and near, kind and cruel, prince and peasant, rich and poor, love and hate, are instances of the innumerable host of things, qualities, and acts that are opposites in character and, consequently, tend to awaken each other in memory whether perceived or conceived. Thus contrary extremes play a considerable part in our ordinary thinking. Recall the notion of a giant and that of a dwarf is likely to follow. A perfectly clear sky begets the thought of a clouded one. The idea of war brings to mind also the idea of peace as its counter condition. The extent of this principle of association will be seen when we consider the fact, that every object of experience, whether it be a thing, thought, quality or act has its opposite. Thus, justice and injustice; freedom, slavery; hope, despair; long, short; mortal, immortal; true, false; good, bad; beautiful, ugly; sweet, sour: are examples of the vast multitude of antagonistic notions that enter more or less into our daily thinking.

106. Every Intellectual Act a Spontaneity in the Beginning.—Every faculty of the human mind is capable of spontaneous action. In fact, a large share of the movements of an ordinary intellect, are simply spontaneities or actions incited by the presence of their objects, without any interference of the will. The beginning, at any rate, of every intellectual act, is always spontaneous and may continue such throughout. In every case the will only deepens and directs a movement that has already begun automatically. If I open my eyes in this room, I can scarcely forbear seeing the various articles of furniture it contains. They present themselves, so to speak, and their images are painted on the retina by no effort of mine. The initial perception that results, is purely a spontaneous one and may remain so to the end. Such perceptions, however, are generally modified by subtle interventions of the will which are more or less strenuous. For instance, I may fasten my attention upon one article to the exclusion of the others, and examine it with special earnestness.

Suppose some one has, without my knowledge, hung a new painting upon the wall, it will inevitably attract the eye. The instantaneous perception that follows, though intensified by surprise, is wholly spontaneous, while the subsequent scrutiny I give the painting, is a voluntary effort. Now this, if I mistake not, is true of every conscious intellectual act. It commences involuntarily on the self-presentment of its object or objects. Of several acts so initiated, the will may select, intensify, and guide any single act, but cannot inaugurate it.

107. All Unconscious Modifications of Mind Spontaneous.—All movements of the possible knowledge which is held by the retentive capacity of memory, are spontaneous. The process by which actual knowledge becomes possible or possible knowledge becomes actual, is an involuntary one. In other words, the final act in committing an idea to memory and the initial act that restores it to consciousness, are both pure spontaneities. An effort of the will is a conscious act which is exerted upon a thought or idea that is now before the mind in the form of conscious knowledge. Volition being itself a conscious act, cannot reach down into the realm of unconsciousness, either to deposit or to awaken ideas. If then the initial movement of an idea when it rises to consciousness in an act of recollection, be spontaneous, the question is, In what manner does the will control the current of ideas which is constantly flowing spontaneously from memory into actual recognition?

108. How the Will Controls the Thoughts in Voluntary Recollection.—It is manifest from facts already presented, that a solitary thought is impossible. The knowledge which is stored in unconscious memory, consists of ideas which are arranged in groups wherein every individual is connected with all the others, by one or more of the associations explained above. Moreover, the groups themselves are connected with each other by similar relations. The special mode of association by which each element of a group, as well as the group itself, is retained and recalled, will, of course, be identical with that under which it was acquired. If we observe, for example, any given fact under the relations of time and place only, we shall inevitably retain and reproduce it under the same relations. Our habits of thought are, therefore, determined not only by the kind of knowledge we acquire, but by its particular mode of association.

The groups of ideas thus held in memory, form what may be called an irregular train which, maintaining a spontaneous activity, passes incessantly under the eye of consciousness. Now the will may select and detain any idea of the train in transit, under conscious notice, and concentrate the attention of the mind upon it. The will not only holds this idea for special scrutiny, but makes it the means by which it so controls and guides the spontaneous train to which it belongs as to bring into consciousness other connected ideas which the mind desires specially to consider. This it does by scrutinizing exclusively that particular quality of the idea under consciousness, by which it is associated with the ideas that are wanted. For we are generally aware that the ideas which any emergency of thought demands, lie in some special line of association whose spring we are, in this way, able to touch.

109. Examples of Controlling Thought by the Will.—Any brief interval of thoughtful experience will supply examples of the operation of the will as modifying the train of thought and determining what ideas it shall reproduce.

Let us suppose at this moment, the current of ideas flowing spontaneously across the field of my mental vision is composed of the names of historic personages and events. Among these the word, Napoleon, arrests my attention. The sign and the thing signified occurring simultaneously, I concentrate my whole mental force on the concept of Napoleon, the Great, and so give it the utmost vivacity. It is manifest that, at my own option, I may scrutinize this concept, as an effect, as a cause, as a resemblance, as a mere local association, or as a contrast. If I concentrate my attention upon Napoleon as an effect, the French Revolution and the bloody period that produced him, come instantly into consciousness. If I think intently upon Napoleon as a cause, then the marvellous changes which his wars wrought upon the nations of Europe, are at once before the mind. If I fasten my attention upon him as a resemblance, there flash into consciousness such characters in history as Cæsar, Cyrus, Alexander and Frederick the Great. Considered as a merely local association, the concept brings into mental view the palace and garden of the Tuileries, the river Seine, and the city of Paris. Regarded simply as a contrast, I shall be apt to compare Napoleon the Great with Napoleon, his nephew, whom Victor Hugo calls " The Little."

Again, pronounce the words " Plymouth Rock," and to a stolid uncultured mind, it will probably suggest only the vague image of a rock and its local surroundings. To the intelligent listener, however, it will, along with the local image, at once recall a historical event with which Plymouth Rock is associated, namely, the landing of the Pilgrims. Fixing the attention upon this event as an effect, reproduces spontaneously the thought of the religious persecutions that preceded and caused it. Pondered as a cause, it calls up vividly the images that represent the result that followed, in settling and civilizing the New World. Considered in the line of resemblance, it will, of course, restore to consciousness events in the world's history that are similar. Further, each one of the concepts recalled in this manner, may, by the selection of the will, be itself made the centre, in which the several lines of association concentrate and bring in their different images at the pleasure of the thinker.

110. Contrast in the Kinds of Thinking.—In the light of these facts, we may easily note the difference between the course of ordinary desultory thought and the more rigid processes of accurate thinking. The one employs without strenuous effort the associations of time and place; the other, the stricter and less volatile relations of cause, effect, and unvarying resemblance. To one man, the sight of a granite bowlder may present only its size, shape, color, location and connection with surrounding objects; in which case, it must be retained and recollected by these local associations only. To another mind, a granite bowlder may not only furnish the facts that attend its locality, but in addition, suggest its constituents of quartz, feldspar, and mica; its origin and classification as an igneous rock; the forces that detached it from its original bed, and deposited it where it lies; and the glacial friction that reduced it to the form of a bowlder. These added suggestions, it is evident, will give breadth and precision to the connections by which the object under scrutiny is afterward held in memory and recalled to consciousness.

Since the associations of time and place are more facile and less severe than those of cause, effect, etc., etc., they become the main avenues of thought, when the mind is in a desultory condition, while, to the ignorant and stolid, they are the principal resources. Perhaps no better illustration of this can be given than the rambling answers of Hostess Quickly in Henry IV.

Falstaff.—" What is the gross sum that I owe thee?"

Hostess.—" Marry, if thou wert an honest man, thyself and the money too. Thou did'st swear to me on a parcel-gilt goblet, sitting in my Dolphin chamber, at the round table, by a sea coal fire, upon Wednesday in Whitsun week, when the Prince broke thy head for likening his father to a singing man of Windsor: thou did'st swear to me then, as I was washing thy wounds, to marry me and make me my lady, thy wife. Canst thou deny it? Did not goodwife Keech, the butcher's wife, come in then and call me Gossip Quickly?

coming in to borrow a mess of vinegar; telling us she had a good dish of prawns, whereby thou did'st desire to eat some, whereby I told thee they were ill for a green wound? And did'st thou not, when she was gone down stairs, desire me to be no more so familiar with such poor people, saying that ere long they should call me Madam? And did'st thou not kiss me and bid me fetch thee thirty shillings? I put thee now to thy book-oath; deny it if thou canst."

QUESTIONS ON CHAPTER V.

Memory second in the order of its growth. Varieties of memory in different minds remarkable. Examples of great men who are deficient in verbal memory, and of others who possess it in a remarkable degree. Examples of men who are deficient in culture, but whose memories are remarkable. Memory strong in the line of our vigorous activities. Examples. Relative values in the varieties of memory. Memory responds to the demands made upon it. What three acts does memory include? Give the order of their action, with examples. Dependence for its vividness, of any act of memory upon a preceding act. Give examples. The initial acts which the senses put forth the most important. What is the first requisite to thorough acquisition by memory? Give examples. The importance of selection in acquiring facts. Promiscuous action of the senses harmful to memory. Give examples. The value of scientific scrutiny to memory. The value of special inclination, or bent of mind, to facility of acquisition. The effect of beauty in stimulating attention. The effect of the music of measure on facility of commitment. Examples of objects whose beauty attracts the attention necessary to acquisition. Give the effect of novelty in stimulating the act of acquiring. What is the effect of strong feelings in stimulating the act of acquisition? What is the character of retention in memory? Define possible knowledge as distinguished from actual knowledge. Give example in Arithmetic. Distinguish between a concept and a percept. Tendency of possible knowlledge to fade. What kind of knowledge soon passes beyond the mind's power of reproduction? What kind of knowledge does memory retain as a permanent possession? By what means is possible knowledge permanently retained? Give the two methods of renewal. Which of these methods is more frequently used in early life? Value of reviews in education. Value of the second means of preserving possible knowledge. Give examples. Retention the storehouse of all our experiences, gathering the concepts produced by every faculty. Recollection depends for its readiness on completeness of acquisition and retention. The laws of association the means by which all knowledge is retained and recalled. What are the principal laws of association? Explain the associations of time, and give example. Point out the associations of time in the quotation from Longfellow's Kavanagh. Describe the associations of place, and give examples. Give the comparative number of our associations of place. What is the frequency of associations in time and place as united in the mind? Point out the associations of time and place in the extract from "White Wings." Also in the extract from "The Lady of the Lake." Explain and illustrate the association of whole and parts. Also the association of sign and thing signified. Subtlety and rapidity of this last association illustrated. Sign and thing signified the association by which we learn language. Differences in the effect of learning language from the written page, and from ordinary conversation. Explain and illustrate the association of cause and effect. Cause and effect one of the relations on which the facts of science are classified. The association of resemblance, its extent and importance. Give illustrations. This association the basis of classification in general. Unvarying resemblance as the basis of scientific classifications. What is the character of association by contrast? Give examples. All our faculties capable of spontaneous action. What is a spontaneity? Every intellectual act spontaneous in the beginning. Give examples. All knowledge held in memory spontaneous. Initial mental movement always spontaneous. Is a solitary thought ever possible? The knowledge stored in memory lies in groups which are connected by one or more of the laws of association, and which together form what is called a train of thought. The train of thought constantly in transit from unconscious memory to conscious notice. Explain how the will selects any idea from the passing train, and makes it the means of recalling another idea from memory. Give example of the process by which the word Napoleon recalls different ideas by means of different associations. What ideas do the words Plymouth Rock recall through its different associations with the events of history that followed? Difference between ordinary desultory thinking and accurate thinking. Illustrate by the granite bowlder. Relation of time and place to desultory thought. Example from the answers of hostess Quickly in Henry IV.

Chapter VI.

CONCEPTION—REPRESENTING CONCEPTS.

111. Conception Defined and Illustrated.—The faculty which comes next in the order of activity is conception. Conception is the power the mind has of grasping and holding up distinctly and vividly for its own contemplation, an idea recalled from memory. A concept, if concrete and representing an external object, is a resuscitated sense-percept. It will be seen hereafter that a concept which is designated by a common name representing a class of things, is a class-concept. It will be shown clearly, as we advance, that a class-concept is the product of the understanding, for which external or internal perceptions simply furnish the materials.

Let us repeat for perfect precision, that conception as we now consider it, represents consciously, in clear outlines and without modification, the individual notions gained from sense-perception. The concept which is its product, is, therefore, if complete, an exact transcript of the sense-percept neither more or less.

In an art-gallery, I notice Rubens' painting of "King David and the Harp." I examine it exhaustively and thus fulfil the conditions for completeness in the acts of acquiring, retaining and recollecting it, that are to follow. Suppose that, after an interval long or short occupied with other matters, some circumstance recalls this painting to my consciousness. If I am then able to represent distinctly and conceive vividly, the mental picture of "King David and the Harp," as I saw it in the gallery, I perform a complete act of concrete conception. Practically, however, the most vivacious concept seldom contains all the elements of the real object it represents.

112. Concepts may be Modified by Imagination.—Beyond question, the mind is able to modify its individual concepts so that they shall exhibit other elements than those contained in the object they represent. But in such case they are, so far as modified, the products of another faculty which we have yet to consider, namely, that of imagination. The following facts include all the more important characteristics of this faculty and its products.

113. Concepts of Sight, Reverie, etc.—It is remarkable that the clearest and most numerous of all our concepts, are those derived from the sense of sight. This is so conspicuously the case, that ordinary thinking consists mainly of the movements of the mind upon sight-concepts which spring up from memory spontaneously through the associations of time and place. In intervals of mental relaxation, when the will is largely suspended, the intellect is often occupied in musing listlessly upon the visual images that float before the eye of consciousness. In such moments, the mind yields to its spontaneities and amuses itself with gazing sluggishly at the figures of the self-moved panorama, as, emerging from the darkness of unconsciousness, they come within the circle of intellectual vision. This form of mental indulgence, which disciplined thinkers allow themselves only as an occasional healthful recreation, becomes with many persons whose powers are less under control, the habitual condition of mind. It is a sort of intellectual dissipation, to which those whose concepts are vivid and unrestrained, too often yield and thereby render the attainment of subsequent high intellectual culture impossible. But the intellect is most completely absorbed in the desultory current of its visual concepts, in dreaming and reverie and those kindred states wherein its fickle flow is scarcely interrupted by voluntary effort. In such states of mind, volition is well-nigh suspended; thought, feeble and purposeless; and consciousness reduced to a mere passive spectator of the fantastic pictures which the eye has previously gathered and memory fortuitously presents.

114. Sight-concepts Superior in Number and Distinctness.—The numerical superiority of sight-concepts is due to the great range and rapidity of the sense of sight. The ideas gained through vision probably exceed in number those derived from all other sources whatever. The greater comparative distinctness of the concepts of sight, may be attributed to the wonderful accuracy of the eye and the infinitely greater attention we give to visible things than to the objects of hearing and touch. Moreover, sounds and the different modes of resistance are simple, while visible objects are highly complex and the various parts of which each is composed, aid and strengthen association, and thus enable us to retain and recall with clearer outlines, the concepts they furnish.

115. Sight-concepts assumed as Standards.—The greater number and vivacity of sight-concepts constitute the reason, if I mistake not, why an ordinary mind unconsciously assumes them as standards for concepts originating from other faculties. For the same reason, children seek instinctively to endow all their ideas with the qualities of visible things, giving color and shape even to spiritual beings. This tendency, which is especially strong in uncultured minds, undoubtedly produced the pictures of Satan and the angels and disembodied spirits found in early books and collections, and reproduced even now in dramas that represent the thoughts and fancies of olden times. These pictures, of course, are the products of imagination; but the concepts of sight are, nevertheless, made the models after which they were fashioned.

116. Concepts of Touch and Hearing.—We have seen that the concepts derived from touch and hearing, are far less numerous and vivid than those of sight. Assiduous special culture of these senses, however, greatly increases the number and vividness of the ideas they supply to the train of thought. This fact is only an example of the general principle that every faculty of the human mind becomes efficient and accurate in proportion to the training it receives.

Many instances are recorded in which the sense of touch has supplied the loss of sight and become wonderfully facile in detecting those qualities and conditions of matter that are ordinarily revealed to the eye alone. Many of the blind who are educated, read rapidly by contact of the fingers with raised letters, and some can even distinguish the colors of different fabrics by their effect upon texture and temperature. The ear also, in case of those who are blind from birth, and necessarily give a large share of their attention to sounds, reaches a degree of nicety in discernment wholly unattainable to those who see. Some musicians have a clearness of conception in music equal to that of sight. It is said of Beethoven, whose ear had attained a marvellous delicacy and acuteness in the perception of musical sounds, that when deafness finally rendered it impossible that he should hear them, his power of conceiving the various notes suggested by touching the keys of the instrument, made their concepts as distinct in his consciousness as the corresponding percepts had been before his hearing was impaired. It is scarcely an exaggeration to say that music furnished the material for his thoughts, and his imagination produced almost involuntarily new combinations in the art.

117. The Effect of Feeling on Conception.—The effect of feeling in quickening and deepening the act of conception, is a subject of great importance which can barely be touched on here. Apathy paralyzes the conceptive power in proportion to its completeness, while a feeling of interest accelerates and imparts lucidity to our concepts, in the ratio that it rises towards enthusiasm. Passion, which quickens every spontaneous mental movement and vitiates those processes of thought that require deliberation, intensifies the concepts that produce it and magnifies them often far beyond their just proportions.

118. Ideal Presence.—It is a remarkable fact, which is expounded at length by Lord Kames, that under the influence of strong emotion, the mind believes in the actual present existence of the object of its concept. Such beliefs are, however, only momentary, being constantly corrected by the evidence of the senses, which shows that they are groundless. The proof that every vivacious idea is attended by a momentary conviction of the external presence of its original, which the senses constantly dispel, lies in the fact that such conviction becomes firm and persistent, in proportion as the senses are lulled or off their guard. Thus, after the death of a dear friend, one grieves over the death-bed scene as though it were a present reality rather than a memory. It is in vain that reflection represents the agony as past and the sufferer as at rest. In the violence of one's grief, the harrowing event forces itself momentarily, again and again, upon consciousness, as an actual presence. And if it reappears as a dream in sleep, when the senses are wholly dormant, it has then the aspect of a reality as valid as when first presented to the senses as an actual occurrence.

This ideal presence, if we may borrow a term from Lord Kames, depends for the degree in which it approaches completeness, upon two circumstances. One is the intensity of passion excited by the concept with which the mind is absorbed, and the other is the extent to which the senses are either deceived or relax their vigilance. In theatrical plays, the accessions of scenery and dress are employed to momentarily mislead the senses; while the acting, which represents the characteristics of the tragedy, is adapted to excite other passions to which it appeals. . It will be seen, on reflection, that, in every fine art, such as painting, sculpture, fiction or poetry, high excellence depends on the measure of ideal presence it awakens in the spectator or the reader.

119. The Teacher's Concepts should be Distinct.—It is important to the success of those who teach, whether with the tongue or the pen, that the concepts they present to others, should be, in the highest degree, definite and distinct in their own minds. The first requisite in imparting knowledge, is that we should have attained it ourselves with the utmost thoroughness. The highest effect of description is gained by selecting and presenting only those striking characteristics of the concepts from which it may be easily realized by the listener. Particularizing the minor details of an idea to be conveyed to another, dims and beclouds instead of elucidating it.

120. What Constitutes the Power of Description.—Genuine skill in description lies, beyond question, in the power to touch the salient points of a subject and leave all others untouched. Great writers, preachers, and teachers possess the talent which selects the distinctive traits and rejects those of less significance. Who could add anything to the effectiveness of Macaulay's description of the character and person of Lord Byron?—

"In the rank of Lord Byron, in his person, in his understanding, in his character, there was a strange union of opposite extremes. He was born to all that men covet and admire. But, in every one of those eminent advantages which he possessed over others, was mingled something of misery and debasement. He was sprung from a house, ancient, indeed, and noble, but degraded and impoverished by a series of crimes and follies which had attained a scandalous publicity. The kinsmen whom he succeeded, had died poor and, but for merciful judges, would have died on the gallows. The young peer had great intellectual powers, yet there was an unsound part in his mind. He had naturally a generous and feeling heart; but his temper was wayward and irritable. He had a head which statuaries loved to copy and a foot the deformity of which the beggars on the street mimicked."

QUESTIONS ON CHAPTER VI.

Define the faculty of conception. Give the distinction between a single concept and a class-concept. Conception as representing individual notions derived from sense-perception. Give the example of King David and the Harp. Concepts as modified by imagination. From what sense are our clearest concepts derived? Sight-concepts furnish the material of our ordinary thinking. What are the characteristics of revery? Revery, if indulged in to excess, is intellectual dissipation. What is the comparative number of our sight-concepts, and what is their relative distinctness? Effect of the complexity of visible objects on the vividness of conception. Why are sight-concepts assumed as standards for those originating from other faculties? Tendency of uncultured minds to modify other concepts by the concepts of sight as standards. Concepts derived from touch and hearing less numerous and vivid than those of sight. Culture increases these ideas both in number and in vividness. The sense of touch often supplies the loss of sight. The sense of hearing becomes exceedingly accurate in case of the blind. Give examples. The effect of feeling in stimulating conception. What is ideal presence, and what are its causes? The conviction of ideal presence momentary. The conviction persistent- when the senses are off their guard. Example of the effect of ideal presence as produced by thinking of the death of a dear friend. The height of passion and the apathy of the senses the causes of ideal presence. Give examples. Value of distinct concepts to the teacher. Skill in description dependent on distinct concepts. Macaulay's description of Lord Byron.

Chapter VII.

ANALYSIS—INSPECTING CONCEPTS.

121. Analysis Defined and Illustrated.—Analysis is the mental act which immediately succeeds conception, from which it derives its material. In the act of conception the mind holds the idea vividly and distinctly before the eye of consciousness, for the purpose of inspecting minutely, one by one, its properties and parts. It is this inspection of each of the properties and parts of a concept separately from the rest, that constitutes the act of analysis.

I am conscious this moment, for example, of the concept of a particular orange. This concept I am able to scrutinize, as though it were an object of sense. I fasten my attention successively and exclusively on each of its parts, and gain thereby notions of its skin, pulp, juice, seeds, etc. Each one of the parts, under this process, furnishes a new concept which is associated afterwards in memory with the notion of the orange, under the relation of "parts to a whole." Further, every concept of a part is subject to a like inspection as to *its* parts. For instance, the parts of the seed may be exclusively scrutinized and thus distinct notions attained of the coats and the embryo that compose it The notions of parts are, in like manner, retained under the association of "parts to a whole."

Again, the mind fixes its attention in analysis successively and singly on the properties which the concept of the orange contains and in this way acquires a notion of each as a distinct entity. The shape, size, color, weight, flavor and structure, come under distinct observation and are affirmed in such judgments as follow: The orange is round, is large, is yellow, weighs six ounces, is sweet, is solid, etc., etc. The mind thus attains individual concepts of the qualities belonging to the concrete concept of this orange, which are associated thereafter with it as a whole and referred to as the roundness, size, color, weight, etc., of the orange.

The above is only the ordinary analysis which the intellect makes instinctively in its first movement upon a concrete idea derived from perception. This act of analysis becomes, as the years of childhood advance, rapid and spontaneous, being incited by the presence of the concept, as visible objects incite the automatic action of sight.

One of the effects of education is to render the act of analyzing our concepts habitually thorough and exhaustive. An analysis that is careless and desultory, is the characteristic of an untrained intellect. On the other hand, the power to make complete analyses, is the product of discipline. A scientific analysis of the orange greatly increases the number and accuracy of our concepts of its structure and constituent elements. This is true of every like process, be the object analyzed what it may.

122. Difference between Scientific and Ordinary Analysis.—The striking difference of results between ordinary and scientific analyses, is exemplified in systematic botany. A careless observer derives from the buttercup, for instance, a concept that contains for analysis only the parts and properties that are obvious to a superficial attention. A subsequent analysis reveals no more elements than the concept includes. Consequently his notions of parts and properties are limited to form, size, color and general outlines. The botanist, on the contrary, finds on analysis that the buttercup has a calix which is slightly colored, regular in form; its leaves or sepals overlapping one another in the bud. The sepals, of which there are five, are concave and deciduous. There are five petals, yellow, alternating with the sepals and much larger than they. The petals also overlap one another in the bud, are plain, dilate, have contracted bases and are deciduous. The stamens are numerous; their filaments, thread-like; anthers, short, the cells opening longitudinally; pistils, numerous, their heads on a globular receptacle; ovary, compressed, one celled, one ovuled; style, short, the stigma occupying its inner side at the apex; ovule, ascending from the inner angle next the base of the cell.

The concept which one who knows nothing of geology, has of a block of syenite, embraces only the superficial elements that are manifest to a careless observer. Analysis would, therefore, discriminate only such properties as size, form, color; while the analysis of a concept of the same object, acquired by an expert in geology, would disclose notions of the more obvious properties of the rock and, in addition, its constituent elements, quartz, feldspar and hornblende, together with the distinctive characteristics of their crystals.

In the light of this subject, it may be easily seen, how widely the study of science extends the range of our knowledge of common things, and how little we should know, without its help, of the planet on which we live.

123. The Concept Depends on the Percept.—It is equally evident that the number and distinctness of the elements which a strenuous effort of analysis shows the concept to contain, will depend mainly on the number and distinctness of the elements which the act of perception has consciously gathered. For the concept can comprise only those known constituents which have been recognized in the previous percept of which it is the counterpart. In fact, the analysis of the concept of a concrete, external object which is absent, is only a more deliberate repetition of the instinctive analysis made in the act of perceiving it. To state the entire law, the number of distinct elements which a valid analysis of any concept reveals, will be in ratio to the completeness of perception, the faithfulness of memory and the clearness of conception, which precede it. From this law, we may derive, if I mistake not, important data in the science of education, which we will develop hereafter.

124. Analysis of an Object under Scrutiny of the Senses.—When a concept is analyzed while its object is under the scrutiny of the senses, the mental acts that precede the act of analysis, are substantially the same as when such object is absent. The only difference between the two cases is in the element of time. When the analysis of a concept is made in the absence of its external object, there has been a perceptible interval between the act of acquiring and the act of recalling it, during which it was retained in memory. The first two acts, namely those of perception and acquisition; and the last two, those of recollection and conception, were separated by a period of time occupied by retention, which was more or less extended.

But when an analysis of the concept is made while its object is present to the senses, the interval of retention is reduced to a minimum too minute to be appreciated. It nevertheless exists, for no psychological fact is better established than that any mental act is impossible without the intervention of memory. In the act of perception, for instance, memory carries forward the results from one indivisible instant to another. In analysis, the mind cannot concentrate its attention exclusively upon a single quality of a concept, without retaining the rest in memory; otherwise there could be no comparison. It follows, then, that in the analysis of a concept, while the object from which it is gained, is under the scrutiny of the senses, the three acts of memory as really precede the act of conception as when such object is absent, and the senses, consequently, are inactive. In such a case, however, these three acts precede with such subtle rapidity, that the interval between perceiving and conceiving is imperceptible, and the percept and the concept seem fused into one. The great advantage of analyzing a concept while its object is inspected by the senses, is that the concept is continually renewed.

In closing this article, it may be well to observe that the concrete concepts derived from the intellectual senses, by no means comprise all the concepts which are susceptible of analysis. Memory acquires and the mind analyzes every concept of a complex character, whether originating from external or internal perception. The processes of thought, complex states of mind, ideas which are the products of imagination or of classification, are, as will be seen hereafter, subject to this fundamental operation. Only those ideas which, being simple, have neither parts nor properties, are beyond the reach of the analyzing process. The origin and characteristics of such ideas will be explained in a subsequent chapter.

QUESTIONS ON CHAPTER VII.

What faculty immediately succeeds conception and acts upon the concept it produces? What is analysis? Give an example of its operation in distinguishing properties and parts of a concept. Give the terms that express the products of analysis as discriminated in the orange. What is the effect of education upon the automatic action of analysis? What are the characteristics of scientific analysis? Give the difference between the scientific and ordinary analysis of the buttercup. Give the complete analysis of the latter. What is the difference between the scientific and the superficial analysis of a block of syenite? On what does the number of elements in the concept depend? Analysis of an object by the senses identical with the analysis of its concept. What is the one difference between the two cases? Describe the difference of the operations between the analysis of an object under the senses and an analysis of the concept of that object in its absence. Memory present in all mental acts. What concepts are susceptible of analysis? Why cannot the mind analyze concepts that are simple?

Chapter VIII.

ABSTRACTION—GENERALIZING CONCEPTS.

125. The Nature of the Process.—We have seen that analysis is the mental process by which the mind separates in thought the parts and properties of a concrete or complex concept, so as to contemplate each one exclusively. The results of this act, when completed, are distinct concepts of the constituent elements of the object under scrutiny. In obtaining these new concepts from the act of analysis simply, the mind does not abstract, i.e., withdraw, them from the concrete concepts in which they cohere. It only concentrates its attention on each one, separately from the rest, of the elements of this particular idea and of no other, the color, size, weight, shape, texture, of this object and this only. Now if I were unable to extend these notions of individual qualities, by comparing them with the notions of like qualities obtained from the analysis of other concrete concepts, I should be far less intelligent than a primitive savage. My mind would be limited to the perpetual contemplation of individual objects without perceiving the relations or the differences between them. But I can perceive not only that this orange which I have analyzed, produces a particular effect on the eye, but that other oranges produce an effect that is identical. Finding, in my notion of an orange a quality called *yellow*, I can take note of it in the notion of another orange and by comparing determine that they are the same. I can find, by further acts of comparison, the identical yellow in a hundred oranges, in many other things besides, and, in this way, come finally to contemplate the notion of *yellow* apart from the things in which it resides, as an abstract quality. By a similar mental process I abstract the notions, red, green, blue, purple, etc., from particular concepts under analysis, hold them up to consciousness and consign them to memory, as abstract concepts.

126. Early Abstraction.—In the primitive efforts of the mind, the child discovers, from repeated impressions on the eye and the instinctive analyses which follow, that a solid has extension in three directions. Such knowledge is certainly vague and indistinct in its early stages, but the process by which it is obtained at last, is always the same. It must be gained,

1. By noticing successively many solids; 2. by analyzing the notions obtained from these solids; and 3. by comparing the obvious properties found in each as existing in all.

By these mental steps and these only, does the child reach, at last, the abstract concepts of length, breadth and thickness. The abstract concepts of shape and size have a like origin in the mind. These are abstracted by comparing them as discerned in the analysis of individual bodies, are found to be the properties of all and, being gradually abstracted therefrom, become finally general abstract concepts.

127. All Abstract Ideas have a Like Origin.—The immense number of such ideas, which every cultured mind has gathered, have a kindred birth. Take, for example, the abstract concept designated by the word *beauty*. In the outset one sees perhaps a rose. Among the characteristics observed in the ideas it furnishes the mind, is that of the feeling of pleasure it gives the beholder. Other flowers, on being successively contemplated, are found to possess a characteristic which produces a like effect. Still other objects, a tree, a lawn, a landscape, a dwelling, a painting, a face, a heroic action; all agree with the flower in the fact that in concurrence they beget in the mind the same emotion of refined pleasure. From these particular objects and the concepts which they furnish by analysis of the property described, the mind abstracts the wide idea which it is able to conceive separately from them all, namely, the idea of beauty. Such is the source of our numerous abstract concepts, usually called abstract ideas.

The origin of abstract concepts of mental attributes, such as honesty, goodness, love, kindness, sincerity, and truth, is, in all respects, similar to that of material abstract qualities. They must each be observed as special instances, before they can be abstracted. For example, we observe many particular instances of states of mind in which men " rejoice with those who do rejoice; and weep with those who weep." Analyzing and comparing these states with each other, we find that they all agree in containing the element of fellow-feeling. Viewing this quality, when abstracted, apart from these special manifestations, we obtain the abstract idea called *sympathy*.

128. Use of Abstract Concepts.—Now these abstract concepts serve two important purposes in subsequent mental operations, namely; they are the units of combination by the use of which the imagination forms new concrete concepts at pleasure; and they are also the standards of comparison which the mind employs in the act of classification.

QUESTIONS ON CHAPTER VIII.

What are the products of an act of analysis? In what does the act of abstraction consist? Illustrate by analyzing and abstracting the concepts of the properties of an orange. Show how the act of abstraction is a series of comparisons. How does the child discover the abstract properties, length, breadth, and thickness, shape and size? In what manner do we gain the abstract concept called *beauty?* Explain how such abstract concepts as honesty, goodness, love, etc., are gained by the mind. What are the two purposes in mental operations that are served by abstract concepts?

Chapter IX.

IMAGINATION.—BUILDING CONCEPTS.

129. Concrete Synthesis.—Since it is the province of imagination to reunite and remodel the separate properties and parts obtained from analysis and abstraction, this mental power properly comes next in order. Imagination is the faculty of mind that combines, at pleasure, the materials supplied by analysis and abstraction into new concrete concepts that have no external counterparts. It is, therefore, a synthetic process.

The centaur was a fabulous monster with the body of a horse and the head and shoulders of a man. This fantastic creature was the product of imagination in the mythological period. A crude analysis of the concepts of a man and of a horse, furnished the parts which were joined together and endowed with life by the imagination.

In one of the galleries of Middle Germany is a painting, called " The Sybil," which represents a woman life-size. The form, expression, attitude, drapery, etc., are suitable to the conception of such a character. Now the painter had previously gathered from observation notions of the more beautiful female forms, and, having analyzed these notions, selected therefrom those that were suitable to his purpose, and combined them into the image concept strikingly represented in " The Sybil."

130. Imagination not the Originator of its own Materials.—A similar examination of any whole which imagination has constructed, will show that its elements have a like source. The imagination does not originate the elements it uses. It is a creator only so far as it works up new wholes out of the materials which the preceding faculties supply. In the average mind, it is a vivacious faculty with a wide margin of spontaneous activity. Its products are, consequently, very numerous—probably far outnumbering the concepts of single things actually existing. Its creations may be designated as image concepts, and distinguished from the concrete concepts of sense in the following characteristics.

131. Distinctions between the Sense Concept and the Image Concept.—The sense concept originates from without; the image concept from within the mind. The first is determined, as to its contents, by external realities. The contents of the second are determined by the mind's own choice. The constituents of the sense concept must pre-exist in actual things; the constituents of the image concept must pre-exist in the sense concept. The purpose of the latter is to realize in thought the actual only; the purpose of the former is to realize and embody the mind's own ideal of the novel and the beautiful, the true and the good. By virtue of the sense concept, man's mind merely duplicates the external world; by virtue of the image concept, he creates a world of his own in which he lives, often in his waking hours, always in his dreams. A well-regulated mind distinguishes accurately between these two classes of ideas by the recollection of their origin, and never affirms that they are identical. For the substance of falsehood lies in asserting that one is the other; that a fact is a fancy, or that a fancy is a fact.

132. Points of Resemblance between Image and Sense Concepts.—Let us note, in the next place, the particulars in which these two concepts resemble each other. Both are held up to the eye of consciousness by the conceptive power; both are alike individual and concrete; both are stored in memory and subject to the same laws of association.

133. Imagination often distorts Facts.—The extent to which imagination modifies or colors, in many minds, the ideas representing facts, and so influencing conviction and conduct, can hardly be overestimated. A person whose imagination is active, but whose habits of observation are careless, often completes the defective concepts acquired through perception by adding thereto imaginary elements. Hence it may come to pass that all his ideas represent only half truths, and are out of joint with the world he lives in.

134. Effect of Passion on Imagination.—The effect of passion or elevated states of feeling is to stimulate the imagination to unwonted activity. Under such incentives imagination not only multiplies its own images, but intensifies and distorts the concepts of sense, until they are no longer the symbols of actual things. When infuriated by passion men not unfrequently lose wholly the power to discriminate between the concepts of fact and the concepts of fancy, and, consequently, judgment and reason become fallacious. In dreams we accept the ideas of imagination as outside actualities; the insane sometimes do this when awake.

135. The Eye furnishes the Principal Materials for Imagination.—Since the imagination is more active and facile in forming concepts of imaginary visible objects than any other products, the separated parts of sight concepts are its most available and abundant materials. Especially is this true of young children, whose ideas are gained mainly from visible things, and whose imaginations are limited, in like measure, for their materials, to parts of sight concepts obtained by imperfect analyses. The child's toys, which for this reason are generally objects of sight, stimulate his nascent creative power to natural action, and realize for him the images of his juvenile fancy. This is shown in the readiness with which his imagination adds to his playthings the lacking elements that make them, for the moment, the creatures they represent. The hobby, the doll, and the mimic man are endowed with life; and the boy at play, spite of all incongruities, even turns into a furious charger the stick he strides. When the educator knows how to train and systematize and apply the true materials for this tendency to spontaneous creation inherent in every child's nature, his processes will be more genuine than they generally are at present.

136. Wide Range of Imagination—Expression. —Writers on this subject have, it seems to me, failed to set forth the wide range and extreme versatility of this faculty, which, more than all others, distinguishes man from the brute. Though its images derived from visible objects, are the more conspicuous ones, yet all the knowledge the mind has gathered, from whatever sources, contributes to its materials and its products pervade the whole territory of human thought. In fact, human activity consists in efforts to express in words or deeds, or to embody in outward forms, the ideas that it originates. The terms, plan, project, scheme, device, expedient, models, to whatever department of human endeavor they apply, are only the synonyms of image concepts. In art, in industry, in every undertaking small or great, the imagination prefigures what the hand performs. The inflexible order of human effort is first the thought and then the thing that duplicates it. The commanding general plans the campaign in imagination, before a column is moved or a battle fought. The thief pictures the stolen treasure in advance of the overt act. Whatever be the moral character of the mental combinations we make, we are constantly striving to clothe them in material forms, and all productive labor, of the hand, the eye, the muscle or the brain, is simply an effort to transmute the inner thoughts into outer facts.

137. Influence of Imagination on Character.— Thus the imagination is a perpetual incentive to action; its images naturally seek expression in language or acts or things, and the unrealized pictures or ideals which our own minds originate, have a powerful influence on our manners, conduct, habits and character.

138. Imagination in the Arts.— It remains to explain briefly, how the imagination operates in forming the original ideas which, when expressed in language or embodied in matter, constitute the productions of art. Of the arts, there are two great classes, namely, the fine arts and the useful arts. The useful arts are those whose products are fitted to supply our ordinary wants and contribute to our comfort and convenience. The fine arts include all those products which are adapted to excite the emotions of beauty or sublimity. Many branches of art, while the central idea they embody is that of use, are closely related to the fine arts, because the structures they produce, receive more or less embellishment. Such an art is architecture.

139. Imagination in the Useful Arts.— Let it be supposed that a machinist is required to fit up a shop with machinery for iron working. The various machines being already constructed, his imagination is limited to the plan he forms for their arrangement. But each machine represents the original idea of its inventor; an idea which his imagination had formed out of materials of kindred character, previously gathered by experience. The draft he makes previous to construction, expresses his concept by lines; but the machine itself, when completed, is its exact transcript. The finished machine which represents an image concept, furnishes subsequently to all who scrutinize it, a sense concept. This is the origin of all the material structures which the useful arts have supplied to the world. Churches, palaces, articles of furniture, the implements of industry, the countless appliances for convenience, are simply the outward semblance of mental structures previously completed. Are the objects of art the only creations that have a spiritual origin? Did not the universe itself pre-exist ideally in the mind of the Creator?

140. Imagination in the Fine Arts.—We have already noticed that, in contributing to the fine arts, the imagination makes those ideal combinations that excite the emotion of beauty. Suppose the landscape gardener desires to range and modify the natural features of a piece of ground so that it shall produce on the eye the highest effect of a beautiful landscape. It is necessary that he should first form a complete picture in his own mind which shall include, in all their details, the improvements he purposes to make, and their combined effects on the eye. He must represent to himself completely the artistic landscape he wishes to make out of a natural one. For this purpose, he has gathered abundant materials from the study of different landscapes, both in their natural condition, and when improved by art. He has scrutinized each element of natural scenery and learned, not only what constitutes the highest beauty in each, but what combination of these elements makes the happiest impression on the beholder.

In selecting his ground, he is careful that its surface lines shall realize his notions of beauty in surface, and he adds thereto in imagination the perfect lawn, the gently curving walks and drives bordered with shrubbery and groups of trees also arranged so as to represent nature in her happiest mood. Having finish his ideal picture, the landscape artist embodies it in a draft which he commits to the care of skilled workmen, who carry it out in all the details of road building, lawn making, planting, etc., etc. The work of the artist lies in the ideal combination, not in the execution of his plans.

Now the landscape painter who is not a mere copyist, gathers the materials out of which he forms his concept of a beautiful landscape, by a process not dissimilar to that of the landscape gardener. He studies nature, however, on a broader scale and his mental pictures represent nature as making her own striking combinations. But the landscape gardener and the landscape painter differ from each other, not so much in their method of forming the idea as in the method of expressing it. For the one employs for this purpose the natural objects which he has ideally combined, while the other represents these objects by means of colors.

In like manner, the musician who engages in musical composition, gathers the elementary concepts of musical sounds, which he has gained through the ear, into new mental combinations which are inspired by elevated emotions of beauty, sublimity or devotion. He has learned from enthusiastic observations, that every passion and every emotion of the human heart has its own peculiar vocal utterance, and the complex concepts which represent his composition are such an adjustment of these elements, as, when expressed in music, are fitted to excite to their highest pitch, the feelings they address. Of such a character, are the compositions of Mozart, Beethoven and Wagner. Thus a musician is great only by reason of his power of imagination. Those performers who interpret his ideas to the world most effectively, are only great singers or great players.

Poetry is ordinarily defined to be a mode of expression, a form of metrical language, appealing to an excited imagination. The truth is, however, that poetry is the product of the imagination, when inspired by the higher feelings, in which a sense of beauty is predominant. Beyond question, the harmonies of metrical language enhance the effect of poetical ideas. But the essential elements of poetry exist in the idea and not in the mode of expressing it. The range of materials for poetic combination, is wide and various. They embrace all that is elevating or heroic in human character and conduct; all that is beautiful or grand in external nature. When poetry deals with the products of mind exclusively, it is called subjective poetry, but when it presents only objects of external nature it is termed objective.

141. Genius.—Genius is an (abundant) power of producing in any department of human thought, new creations which, when expressed, shall accord with the decisions of a cultivated judgment. Thus Napoleon was a genius in war; Newton, in science; Shakespere, in dramatic poetry; Dickens, in fiction; Darwin, in biology; Rubens, in painting; Bell, in architecture.

142. Taste.—Taste is properly limited to the fine

arts. It is simply an accurate judgment as to the fitness of the ideas of the imagination to produce the emotion of beauty. Genius, then, is a characteristic of the creative faculty possessed by few. Taste is the arbiter which pronounces upon the value of the creations of genius in the fine arts. It is fortunate for the world when genius and taste are united.

143. Language.—Words are the artificial signs by means of which one mind communicates its ideas to another. The language of significant sounds which makes intercourse possible, sprang from the universal desire of man to share his ideas and feelings with his fellows; in other words, language is the outgrowth of human sympathy and, without its aid, the human mind, as will be shown hereafter, could never get beyond the crudest processes of thought.

My present purpose is, however, to show the effect on our image concepts, of clothing them in words. In studying the principles of association, we learn that the mental train presents to the eye of consciousness, successive clusters of ideas. Now, if the mind selects and designates with a name any of the ideas of a cluster, they are generally enlivened and intensified thereby. All the various concepts which the different operations of thought produce, are rendered more clear and conspicuous to consciousness, by clothing them in words. Especially is this true of the image concept. The specific expression of a concept of the imagination, gives it a peculiar vividness and distinctness of outline. This is the reflexive effect of words on the mind of him who utters them. On the other hand, the effect on the listener is to stimulate his imagination to form the picture the word suggests. Thus when reading a poem, we are constantly striving to realize, by an act of the imagination, the images the language depicts.

The images presented in the following passage, furnish the incentive for such an effort.

> " He who ascends to mountain top shall find
> The loftiest peaks most wrapped in clouds and snow ;
> He who surpasses or subdues mankind,
> Must look down on the hate of those below."

When a word, written or spoken, designates a concept which the mind has previously gained it recalls such a concept from memory; but when the word signifies an object which is outside of personal experience, the notion of such an object is realized through imagination. If, for instance, we speak of the cathedral of Cologne to one who has personally visited it, it recalls a concept from his memory, but when the dome of St. Peter's, which he has not visited, is mentioned, his imagination constructs an image of it, out of materials gained by observation of like structures. The image formed to represent St. Peters, will depend, for its completeness, on the vigor of the imagination and the abundance of its appropriate materials.

QUESTIONS ON CHAPTER IX.

What is the faculty that naturally stands next in order? Define imagination, and give examples of its operation. Imagination not the originator of its own materials. Comparative number of the concepts which imagination has produced. Give the characteristic distinctions between the sense concept and the image concept. What peculiar purpose does each subserve? What are the points of resemblance between sense concepts and image concepts? The tendency of imagination to modify sense concepts. The effect of passion or strong feeling upon imagination. Imagination in dreams. Which of the senses furnishes the principal materials for imagination? Why are children's toys generally objects of sight, and what is their effect on imagination? Materials for imagination gathered from all sources. What are the various means by which imagination expresses its concepts? Human action consists largely in efforts to express the products of imagination. What is the influence of imagination upon character? What is the relation of the imagination to the arts? What are the two great classes of arts? Explain and illustrate the operation of imagination in the useful arts. Give examples. What is the relation of imagination to the fine arts? Illustrate by the operations of a landscape gardener. Also by a landscape painter. What is the office of imagination in musical composition? Poetry a product of imagination. What are the essential elements of poetry? What is the effect of metrical language on poetry? In what does genius consist? Define taste, and explain its relation to the arts. Of what feeling is language the outgrowth? What is the effect on our image concepts of clothing them in words? What is the effect of such words on the listener? Under what conditions is the meaning of a word realized through imagination?

Chapter X.

CLASSIFICATION—ARRANGING CONCEPTS.

144. The Mental Acts which Precede Classification.—In the special acts of sense-perception, memory, and conception, we regarded the mind as gaining, retaining and representing to consciousness a concrete concept which was no more than a transcript of the external object from which it was derived. We considered the three powers referred to as consecutively employed in gathering single ideas from particular things with which the intellectual senses come in contact.

Now, if the human intellect were limited in its action to representing the individual objects presented to sense-perception, our minds would simply reflect like a mirror the miscellaneous things around us, and the world within would simply duplicate the world without. If such were the condition of the human mind, memory would be overwhelmed with a vast throng of particulars and a countless multitude of proper names necessary to express them. But the facts are far otherwise. In large part, the ideas with which the memory is actually stored, represent not single things but classes of things; and the words we use, designate mainly not individual concepts but groups of concepts which represent classes. As we have already seen, the mind has the power to modify its single concrete concepts gained from without, by the several processes described.

1. It analyzes these concrete concepts and obtains therefrom concepts of the parts and properties they contain.

2. It abstracts, by comparison, these individual properties and obtains therefrom abstract concepts of properties apart from the concrete concept that contained them.

3. It combines, at pleasure, the concepts of parts and abstract properties into new concrete image concepts having no external realities.

4. It arranges, by a process I am now to explain, its single concrete concepts into classes, each of which is designated by a single name. The words, *house, quadruped, farm,* for instance, embrace each a group of objects that resemble each other in certain uniform characteristics. The noun, *house,* gathers under its significance all the single structures built for human habitation. Observe that, however much the single structures may differ as to form, size, color, and the materials of construction, they all agree in the purpose for which they were built. Thus, the one characteristic *of purpose* in which they resemble each other, constitutes the ground on which they are united in a separate class and distinguishes them from all other structures. Now, since a definition is a formal statement of the ground on which the thing defined is assigned a distinct class: *A house is a structure built for human habitation.*

Again the word, *farm,* applies to pieces of land of various areas which, though differing in many particulars, must agree in the characteristic on which the class is founded, namely, that they are devoted to the raising of agricultural products. This essential circumstance in which all farms resemble each other, is the basis of the class and distinguishes the farm from all other tracts of land; consequently, *a farm is a piece of land set apart for the raising of agricultural products*.

Further, the word, *quadruped*, covers with its meaning, a multitude of animals which, disagreeing in a great number of particulars, all resemble each other in the essential characteristic that they have four feet. *Hence a quadruped is an animal having four feet.*

Considering next the noun, *river*, we find it includes all large streams of water which flow in channels into other bodies of water. Since all streams on the earth's surface, great and small, flow in channels into other bodies of water, the single property on which the streams covered by the word river, differ from all others and resemble each other, is superior size.

These simple examples serve to show, that the class names of which language is largely composed, designate groups or classes of things which resemble each other in one or more uniform properties or characteristics that bind them together and, at the same time, exclude all other things.

145. The Mental Process of Classification.— Consider next the mental process by which the act of combining concrete concepts into classes, proceeds. The child, whose powers of thought are gradually unfolding, must classify his early ideas by the same successive operations that the mature mind performs much more perfectly and rapidly. Playing daily with the household dog, for example, he attains by many repeated acts of sight, hearing and touch, a concept of the animal as a whole. Even before it is completed, his mind is making, by similar repetitions, a crude analysis by which he gets elementary concepts of parts, as heads, tail, body, legs, mouth, teeth, hair, etc., etc., and also of obvious properties and characteristics; as form and size, whole and parts, relation of parts to each other; likewise of voice and actions. At the sight of another like animal, the child, by successive repetitions of spontaneous perception, 1. Gets a notion of it as a whole; 2. Analyzes this notion into its parts and properties; 3. Compares the notions of these with the notions gained from its first analysis, and finds by incipient abstraction that they are similar; 4. Classes the two together and calls the second animal by the same name, dog.

Now these movements of the infant mind in its first attempts at classification, are far more vague and imperfect than a formal statement of them would imply. The feeble successive efforts are divided by no rigid lines. They overlap and intermingle, yet their order is essentially fixed. The child must attain a concept of an object, as a whole, before he can analyze such concept into parts and properties; must compare the individual notions of parts and properties thus attained, with like notions attained from analyses of other concrete concepts, before he can get rudimentary abstract concepts; must use these elementary abstract concepts as standards for determining that two or more concrete concepts, having identical properties, stand together in a rudimentary classification.

More briefly, the first dim, imperfect, intellectual steps which result in incipient classes of ideas, may, if we include the initial one, be indicated by the participles, perceiving, remembering, conceiving, analyzing, abstracting, classifying. Beyond question, the genuine processes of early education consist in stimulating each of these operations in its order, to greater effectiveness, by presenting its appropriate materials.

146. Desultory Classification.—We have seen that mental activity consists in a series of consecutive operations which inevitably culminate in classification. These operations in children, savages, and ignorant persons, generally are, from first to last, desultory and imperfect. The incomplete classification that results, is consequently founded on obvious characteristics that the senses gather spontaneously, while points of resemblance more important and less superficial, escape observation. Ignorance reveals itself uniformly by its scant knowledge of the characteristics included in the class concepts. Thus to a person with no other knowledge than that gained by superficial observation, a fish is an animal which lives in the water and has a body furnished with fins and gills.

147. Scientific Classification.—To one whose antecedent observations and analyses have been thorough, a fish is an animal that inhabits the water; has cold red blood; its heart two-chambered; breathes by gills instead of lungs; has fins as means of motion; no external ear; eyes immovable; organs of voice entirely wanting; skin either naked or covered with scales; most prolific of oviparous animals, a single female often producing over a million eggs in a season; longevity great but unknown.

To add another similar example: A desultory classification would contain only the facts that a bird is a biped whose head is furnished with a bill for taking food, and whose body is covered with feathers and supplied with wings for flying. But the elements gathered from scientific observation and analysis and conciously comprised in class concept, would be as follows:

A bird is an animal whose body is covered with feathers and furnished with wings for flying, instead of the fore legs belonging to other animals; has the great muscular power needed in sustained flight; air in the body not confined to the lungs but carried to various large air cells and even to the interior of the bones, which are hollow for that purpose; bill performs both the office of jaws and fore paws in other animals; vocal organs highly developed; plumage, annual moults; digestion powerful, completed in an organ called the gizzard; sight very acute; etc., etc.

148. Perfect Classification.—We have seen that our processes in classification are often incomplete and desultory, whenever the objects classified are concrete concepts acquired through the senses. When, however, the ideas to be arranged in classes, consist of abstract concepts, the case is wholly different. Classes formed by repetition of identical, abstract ideas, are absolutely complete and perfect.

A familiar classification may be found in the decimal system. The notion of number is one of the earliest of those notions which we have abstracted from the concrete ideas of individual objects around us. Taking the abstract concept of a unit, we repeat it until we reach the number ten. Assuming this ten as a new unit, we make the same number of repetitions with it, and reach a hundred. The hundred being now our unit, we repeat the same steps and reach a thousand. The successive units which serve as bases for additional classes, are tens of thousands, hundreds of thousands, millions, etc., etc. It is manifest that this classification would be equally complete and perfect, if the multiplier with which we make, in each step, the complex unit, is one which conception can easily grasp. Suppose, for instance, the system to be quarternary instead of decimal, four simple units would then make the first complex one $1 \times 4 = 4$. The second complex unit would be $4 \times 4 = 16$. The third $16 \times 4 = 64$. The fourth $64 \times 4 = 256$. And so on, indefinitely.

It is manifest that the superiority of the denominations of Federal money over those of other nations, consist in the fact that the units are identical with the units of the decimal system. Its multiplier being 10, the mill, cent, dime, dollar, eagle, etc., are the successive units. Though, theoretically, the mill is the fundamental unit, practically the cent is the real simple. The nickel is half the first complex unit; the twenty-five cent piece is one fourth and the fifty cent piece one half the second complex unit, or the dollar.

The denominations of English money are less convenient, because of the variable multiplier which produces the different complex units, namely, the farthing, the penny, the shilling, and the pound. The same criticism may be made on the other systems whose multipliers are varying numbers. The metric system of weights and measures adopted by France, in which the standard unit has the least possible variation and the multiplier is identical with that of the decimal system, has for this reason, a superior value which will ensure its final adoption by other civilized nations.

It will be readily seen that we are able to count only those things that belong to the same class, and that none but objects which are classed together, can answer to the question, how many? If, for example, in a gathering, there are individuals of both sexes and all ages, we count them as so many people or persons. If such a gathering consists of the young of both sexes we say there is such a number of children. If they are of one sex only, they can be counted as so many boys or girls as the case may be. I can count in this room a dozen things, but in so doing, as they are quite unlike, I am compelled to class them together on a single common characteristic, namely, their use as articles of furniture. A few of these having the same use and form, may, indeed, be enumerated in a narrower class, as six chairs, two tables, etc., etc.

149. Classifications Based on Form.—The abstract ideas of form, as well as those of number, may serve as an instance of complete classification made by the repetition of resembling units.

As an illustration, the words, "rectangular isosceles triangle," include in a single perfect class, a multitude of rectangular triangles having two equal sides. These again are included in a wider group, namely, the rectangular triangle which forms a complete class of individual triangles having a right angle as a common characteristic. The right angle triangles further belong to a still higher class under the single term, triangle, in which the individuals agree in the common fact of having three sides and three angles. But the triangle is a member of the more extensive group, called plane figures, in which we have the square, parallelogram, polygon, circle, etc., etc., in short, any figure bounded by lines on a surface. But the plane figures are embraced in the higher and more extensive class termed regular figures, which include the entire catalogue of geometrical forms, whether plane or solid. Finally, the regular figures are gathered into the vast group which contains all forms that bodies may assume or the mind conceive in space, whether regular or irregular.

We have then, under the word, figure, our widest possible group, which includes all existing or conceivable instances, classed together on the single characteristic of outlines. On the other hand, the lowest group in the series contains the fewest number of individuals classed together on resemblances that are numerous and perfect. A simple diagram will show how classification, beginning with the least number of closely re-

Classification—Arranging Concepts.

sembling objects, rises by steps to final classes which embrace the largest number of objects having the least resemblance that will justify classification.

Figure...
Regular figure..
Plane figure.....   ....
Triangle...
Right angle triangle.....................
Right angles isosceles triangle.......

Here under class 1st, figure embraces all the objects having outlines. The next lower class 2d, under regular figure, contains those figures of class first having outlines that are regular. Class 3d, plane figure, contains only those figures of class second having outlines that are regular, and are drawn on a plane surface. Class 4th, under triangle, groups together exclusively those figures of class third having outlines that are regular, drawn on a plane surface and having three sides and three angles. Class 5th, termed rectangular triangles, extends simply to figures of class fourth, having outlines that are regular, drawn on a plane surface, having three sides and three angles one of which is a right angle. Class 6th, the rectangular isosceles triangle gathers in, lastly, those figures of class fifth having outlines that are regular, drawn on a plane surface, having three sides and three angles, one of which is a right angle formed by two lines which are equal in length.

Observe that the lowest class, the rectangular isosceles triangle, comprises a comparatively small number of objects classed together on six common characteristics. Observe, further, that the highest class, figure, comprehends the largest number of objects united by a single characteristic held in common. Each individual of the lowest class unites in itself all the properties on which the classes above are formed, besides its own peculiar property on which its class is based, namely, a right angle formed by two lines of equal length. Descending from the highest class, figure, at every step, we divide or diminish the number of objects and increase the number of resembling characteristics. Ascending from the lowest class, rectangular isosceles triangle, we, at each succeeding step, increase the number of classified objects and diminish the number of resembling properties.

The number of individuals embraced in the class is called by Sir William Hamilton its *extension*, and the number of resembling properties on which its classification is based, its *comprehension*. It is manifest that extension is at its maximum and comprehension at its minimum in the highest class; while, in the lowest class, comprehension is at its maximum and extension is at its minimum.

Figure.—Having outlines.
Regular Figure.—Having outlines that are regular.
Regular Plane Figure.—Having outlines that are regular and drawn on a surface.
Triangle.—Having outlines that are regular, drawn on a plane surface, and having three sides, and three angles.
Right Angle Triangle.—Having outlines that are regular, drawn on a plane surface, and having three sides and three angles, one of which is a right angle.
Rectangular Isosceles Triangle.—Outlines that are regular, drawn on a plane surface, having three sides and three angles, one of which is a right angle, formed by two lines of equal length.

150. Definition.—From these laws of classification, we may easily learn the nature of a definition. A definition designates a class to which the thing defined belongs, and adds the characteristics which consign it to the class next below; as follows:

A rectangular isosceles is a right angle triangle having two equal sides.

A right angle triangle is a triangle one of whose angles is a right angle.

A triangle is a plane figure having three sides and three angles.

A plane figure is a figure drawn on a plane surface.

A regular figure is a figure having regular outlines.

151. Scientific Classification of Concrete Objects.—The successive steps in natural classification, may be further illustrated by an example from zoology:

Animal..Kingdom.
Vertebrate... Branch.
Bird..Class.
Climber..Order.
Woodpecker..Family.
Ground Woodpecker.................................Genus.
Golden-Winged Woodpecker.......................Species.

The golden winged woodpeckers are classed together on the many common characteristics of form, size, color, food, habit, parentage, that distinguish the species. This species possesses a few organic peculiarities that are similar to those of eleven other species constituting the genus, ground woodpecker. But the ground woodpeckers are, by shape of bill and habit of pecking for their food, members of the great family of woodpeckers. The woodpeckers, from the form of their feet, which are adapted to climbing, belong to the climbers, which constitute one of the four orders into which the class, bird, distinguished by the organs that enable it to fly, is divided. The bird by virtue of possessing a back bone, is a vertebrate, which class again is one of the four great branches of the animal kingdom.

152. Comprehension and Extension.—In this series of classes, the golden winged woodpecker, containing all the characteristics of the groups above, in addition to those which distinguish its species, exhibits the greatest comprehension and the least extension. On the other hand, the class, animal, which crowns the series, embracing an entire natural kingdom, has the largest extension with the least comprehension.

We have thus seen that all knowledge, except the most rudimentary, is arranged and retained in a series of classes, where each class comprises its own peculiar concepts or things, and is itself comprised in the next class above it, until we reach the highest group, in which extension attains its maximum and comprehension its minimum. To know a thing is to understand, accurately and exhaustively, the characteristics of the species and genus to which it belongs, as well as those of the higher classes in which these are successively included. The mind, in the act of thinking, assigns a concrete concept to its proper class, or a class concept to the wider class that includes it.

153. Office of Language in Classification.—It is an interesting fact that language, instead of employing proper names which are comparatively few, generally specifies a particular individual or a narrower class included in a wider one, by means of adjunctive words, phrases, or sentences. In fact, the main purpose of modifiers in a sentence, is to express an idea with exactness by reducing classification from wider to narrower groups or to individuals.

One or two illustrations will show this office of modifiers more clearly. In the clause, "earnest thoughtful men of this city, who abhor injustice," the principal concept is *men*, which, standing alone, embraces a wide class. But the adjective, *thoughtful*, evidently reduces the extent of the class, *men*, to the narrower class, *thoughtful men;* which again is limited still more by the epithet, *earnest*, since it is manifest that the class, *earnest, thoughtful men*, while belonging to the class *thoughtful men*, belongs also to a less numerous class below it. But the group, *earnest, thoughtful men*, is reduced to a still narrower group, by the phrase, *of this city*, and this again to a group comprising a still smaller number, by the adjunctive sentence, *who abhor injustice*.

The series of classes then embraced in the example we have chosen, stand, as follows:

HIGHER CLASS......1. Men.
                          2. Thoughtful men.
                          3. Earnest, thoughtful men.
                          4. Earnest, thoughtful men of this city.
                          5. Earnest, thoughtful men of this city, who abhor injustice.

The same inverse ration of extension and comprehension exists in this series, as in all others; the highest class comprising the largest number of individuals, the lowest the largest number of characteristics.

QUESTIONS ON CHAPTER X.

What powers have we seen to be employed in gathering single ideas from particular things? What would be the result if the intellect were limited to representing individual objects? What is the actual condition of the ideas we hold in memory? What are the three mental operations that precede classification? What ideas do the words we use for the most part designate? What group of objects does the word house denote, and what characteristics does each object in the group have in common with the rest? Define the words, house, farm, quadruped, river. What are the operations by which the child classifies its early ideas? Classification under the name, dog. First movements of the infant mind in classification vague and imperfect. Give the intellectual steps in order that precede classification. Desultory classification by children and savages. Scientific classification as illustrated by a fish. Desultory classification of a bird contrasted with its scientific classification. What are the conditions of perfect classification? What kind of concepts are capable of a perfect arrangement in classes? Example of the decimal system. Superiority of Federal money. Why are the denominations of English money less convenient? Superiority of the metric system of weights and measures. Process of counting impossible without classification. Complete classification in groups as illustrated by mathematical forms. Show how classification makes groups within groups. Write out and explain the classification of mathematical plane figures, containing groups within groups; and give the characteristics on which each group is based. Name the class in which the number of objects contained is least and the number of common characteristics greatest. Name the class which contains the largest number of objects and the least number of common characteristics. Show how the number of objects decreases and the number of resembling characteristics increases as we descend from the highest to the lowest class. Show how the number of objects increases and the number of resembling characteristics decreases as we ascend from the lowest to the highest class. What term expresses the number of individuals embraced in a class? What term the number of characteristics? In what class is extension at its maximum and comprehension at its minimnm? In what class is comprehension at its maximum and extension at its minimum? Write out the characteristics of each of the descending classes under figure. What is definition? Give examples. Define each of the mathematical plane figures. Classify in

Classification—Arranging Concepts.   133

ascending groups the Golden Winged Woodpecker, and show that extension and comprehension increase and diminish inversely. By what expedient does language designate individuals without using proper names? Show how modifiers reduce classes from wider to narrower groups.

Chapter XI.

THE UNIFORM SUCCESSION WHICH PRECEDES CLASS JUDGMENT.

Judgment—Thinking in Concepts.—Hitherto we have confined our attention to those successive acts by which the intellect acquires, retains, recalls, scrutinizes, and classifies its individual concepts. We have found that the individual concept originates in perception, is consigned to memory, is retained and restored to consciousness wherein the mind grasps and concentrates its attention upon it. We have learned that the first of these acts, namely, perception, is a conscious one and that if strenuous and exhaustive, it fulfills completely the conditions for valid acquisition by memory. We have seen that the successive acts of memory which follow perception, namely, those of acquiring, retaining, and recalling, are strictly unconscious and automatic, and that, in the final act of recalling, the mind can control the concepts that emerge from consciousness, only by fixing its gaze upon and quickening a present conscious concept connected therewith by some line of association. It has been made clear that conception which is the crowning act of the series, depends, for its completeness, upon the thoroughness of the preceeding acts, especially, the initial act of perception. The final product of the series, as we have seen, is the concept which is conscious, individual, and concrete. Now it is upon this concept and the material it furnishes, that the intellect performs all its subsequent operations in the act of thinking. For thought centers on the concept, and its various movements consist in dividing such concepts into parts and properties; in separating its qualities and scrutinizing each apart from the rest; in reuniting these qualities in different combinations at pleasure; in grouping concrete concepts into classes on the basis of resembling characteristics; in comparing these classes with each other and drawing therefrom new conclusions respecting the truths they contain these are the processes by which individual concrete concepts are modified by thought and wrought into the higher forms of knowledge. So far, each succeeding chapter has elucidated, step by step, the order of their activity.

155. Comparison the Fundamental Act in Thinking.—Every movement of thought involves a comparison between two concepts. The simplest act of discriminating one concept from another, implies a comparison of the two. Since every concept gained through the senses, represents, so to speak, a bundle of qualities, the mind cannot fasten its attention upon one of these in an act of analysis, without distinguishing it from the other qualities, and this constitutes a simple comparison. Nor can the analysis of a concrete concept into parts, be conducted without comparing these parts with each other. Nor can these parts be recombined into new concepts without continued comparison. It is equally evident, that the act of classification is the grouping together of those concepts whose resembling characteristics are discerned by comparing them together. In short, every act of thinking, whether it be to analyze our concepts or to combine the elements that result from analysis, into new forms, includes inevitably a comparison.

Take such examples of the simplest forms of thought, as:

The sky is blue; the town is not distant.

In the first, the mind compares the two concepts, sky and blue, and affirms that the one belongs to the other. In the second, the mind compares the notion of a town, with that of distance, and denies that the former belongs to the latter. Undoubtedly, these acts of comparison, which are essential to every form of thought, are often so subtle and flitting as to find no lodgment in the memory. The operation of comparison in the presence of familiar concepts, is spontaneous, resembling the automatic action of sight, when the eye rests involuntarily on visible objects.

In the sentence, "Twelve cities were free; the others acknowledged allegiance to the emperor." Since all objects that are grouped together under a number, are classified, the expression, twelve cities, necessitates a comparison previous to grouping. Then again these cities, must have been compared, in order to determine the condition of freedom that is affirmed of them all. In like manner, the word, *others* implies a comparison; first, as to the general resemblance that binds them together as cities; and secondly, as to the common allegiance to the emperor which they acknowledged. Finally, a comparison between allegiance and emperor, is expressed by the preposition, *to,* and it is necessary to compare, *allegiance* and *acknowledged* together, in order to recognize the relation of the act to its object. All these discriminating movements run through our thoughts in such subtle flashes, as to render any explanation of them a comparatively slow and clumsy performance. It is like gathering a broad landscape with a sweep of the eye, and then turning away and trying to reproduce and describe the innumerable details of the picture.

156. Class Judgment in Especial.—Judgment is a mental act which, comparing two concepts, affirms that one is or is not contained in the other. Thus it will be seen that judgment is a comparison, completed and affirmed. Thought proceeds in judgments or, in other words, every complete thought is a judgment or a series of judgments. We think in judgments and there is no other form of complete, conscious, intellectual action. Judging and thinking then are identical acts, differing only in the fact that the latter may consist of a single judgment or several connected ones. We can consequently learn the nature of thought only by studying the variety of judgments of which the mind is capable. For thought, moving always in judgments, is a comparison of concepts or a succession of such comparisons, made in order to discern and affirm the agreement or disagreement of related ideas.

When it is said that, "judgment affirms," a careful distinction should be made between the words, *affirm* and *express*. For the verb, affirms, designates, in this connection, simply the mental act which is essential to a judgment; while the word, express, would signify the clothing of that act in language. The mind affirms in thought but expresses in words. A comparison of two concepts, when affirmed, is a judgment or thought; when expressed, a proposition or sentence.

But let us illustrate the simplest acts of judgment by adequate examples: Take the following propositions:

1. Snow is white.
2. Vinegar is sour.
3. The wind is chilly.
4. The sun is hot.
5. Peter is not a laborer.
6. Thomas is not truthful.
7. I am not a beggar.
8. The traveler was penniless.

In the first, we compare the concept, snow, and the concept, white; and affirm that the former—in extension—contains the latter. That is, the class of objects named snow, is included in the wider class named white.

In the second, vinegar and sour are compared and the first affirmed to be included in the wider class, sour. That is, vinegar is among the things that are sour.

In the same way, the third judgment affirms that wind is contained in the class, chilly; and the fourth that sun is contained in the class, hot. Remembering that snow, white, vinegar, sour, wind, chilly, sun, hot, are class concepts we may represent these judgments by circles.

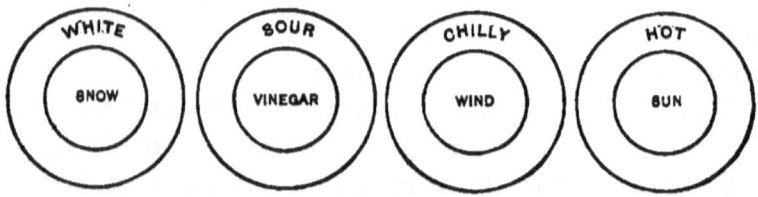

Of the negative judgments, the fifth affirms, that the concrete concept, Peter, is not included in the class concept, laborer; the sixth that the individual,

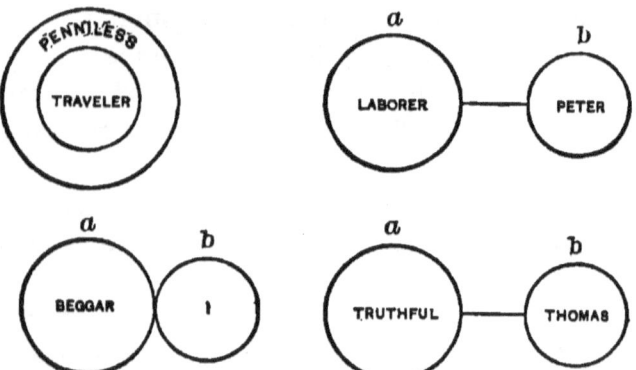

Thomas, is not within the class, truthful; in other words, that Thomas is not among truthful persons;

the seventh excludes the concept designated by the pronoun, I, from the class, beggar; the eighth affirms that the class, traveler, is included in the class, penniless. Like the affirmative judgments, the negative judgments are also the results of completed comparisons, which may be represented as in the lower figures on the opposite page.

157. The Proposition.—Having gained a glimpse of the nature of a judgment, let us turn our attention to the proposition in which it is uniformly expressed, when expressed at all. We have seen that a judgment is the comparison of two concepts, in which one as a part, is perceived to belong to the other as a whole. The proposition which is a judgment embodied in language, is regularly composed of three words, two of which designate the concepts compared and the third, the neuter verb, affirms the relation between them and is termed the copula. The concept of which something is affirmed, is called the subject. That which is affirmed of the subject, is the predicate. In the propositions, the horse is black; time is fleeting; the road is long; the elephant is a quadruped. Horse, time, road and elephant, are each the subject of the proposition in which it stands, while black, fleeting, long, and quadruped, serve as predicates.

In the natural order, the subject of the proposition, stands first, the copula next, and the predicate last. But, in language, this order is frequently inverted, in which case, the subject answers to the question, what is that of which something is affirmed? And the predicate to the question, what is affirmed of the subject?

As will be easily seen, the subject must uniformly be a noun, a pronoun, or another part of speech used as a noun. The copula is always the neuter verb or some equivalent. The predicate may be a noun, pronoun, adjective, participle, phrase, or dependent sentence.

Examples.

1. He is educated.
2. You are wrong.
3. The passage is perilous.
4. To err is human.
5. The prisoner was a knave.
6. The woman is in trouble.
7. Fighting is forbidden.
8. I am he.

Each of the above propositions, except the last, contains a judgment or proposition in which the subject is recognized, as belonging to the wider class embodied in the predicate. This is easily represented, as as follows:

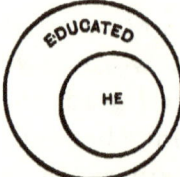 He belongs to the class of educated men.

 You belong to the number of those that are wrong.

 The passage is one of those that are perilous.

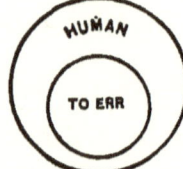 To err is one of the acts of a human being.

Uniform Succession which Precedes Class Judgment. 141

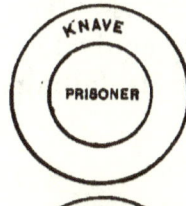 The prisoner is one of the knaves.

 The woman is among those who are in trouble.

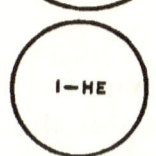 I and he are coextensive and identical.

The subject and the predicate may each be modified by adjuncts to show the complexity of the concepts they embody.

A cloudy sky and a south wind betoken heavy rain.

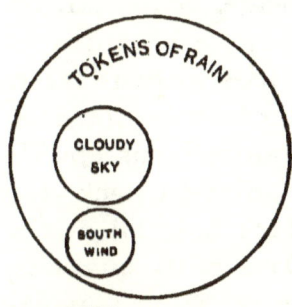 A cloudy sky and a south wind are among the tokens of rain.

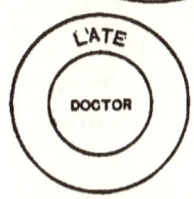 The doctor is probably late

The doctor is probably included in the class, late. Often the predicate and the copula are united in the same verb, as: "other worlds exist." He is. The rain falls. These propositions are evidently equivalent to—other worlds are existing. The rain is falling. He is existent.

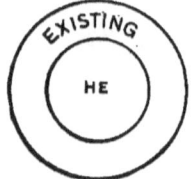

He constitutes one of the things that exist.

158. Judgment in Extension and Comprehension.—So far we have considered the subject in a judgment as contained in the predicate; but Hamilton has clearly shown that, in affirming a judgment, the thinker may at his option consider the subject as a part of the predicate, or the predicate as a part of the subject. For example, in the judgment, "the horse is black." The subject, horse, may be regarded as one of the class designated by the predicate, *black;* or, on the contrary, the subject, horse, may be regarded as containing *black* as one of its qualities without reference to the existence of this color in other objects. The same form of proposition, therefore, is susceptible of these two interpretations: The horse is black: the horse is contained in the class, black objects; or, the horse contains the color, black, as one of its qualities.

In the first case, the predicate is considered as extended; that is, it is regarded as applied to the entire number of things that are black, of which the horse is affirmed to be one; and the judgment is termed a judgment in extension.

In the second case, the predicate, black, is regarded simply as comprehended in the horse as one of its qualities, and the judgment termed a judgment incomprehension. To decide, then, whether a judgment is in extension or in comprehension, we have only to ask, is the predicate considered as a class of which the the subject is one? Or is the predicate considered as a single quality which is a part of the subject? In the first case, we have a judgment in extension; in the last, a judgment in comprehension.

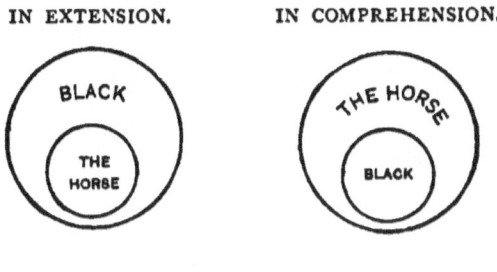

QUESTIONS ON CHAPTER XI.

To what series of operations have we given our attention hitherto? Describe these operations minutely and give their result in the attainment of knowledge. Show the office of comparison in each of the successive steps. Point out the acts of comparison in the simplest sentences. What is the nature of comparison when the concepts compared are familiar? Give examples. Subtle movements of comparison. Define judgment as based on comparison. Thought and judgment identical acts. Distinction between the word, affirm, and the word, express. Give the distinction between a judgment and a proposition. Write several simple propositions, and show that each affirms that a narrower class is included in a wider one. Show the relation of the class containing and the class contained, by two circles. Write several negative judgments, and show by circles how one concept is excluded from another. Analyze a proposition which expresses a judgment, and name and define its parts. What is the natural order of a proposition? Write several propositions and show by circles the relation to each other of the two concepts they express. What is a judgment in extension as distinguished from a judgment in comprehension? Show by a circle within a circle the relation of the two concepts of a judgment in extension. Also, in the same manner, the relation of the two concepts of a judgment in comprehension.

Chapter XII.

REASONING—INFERRING CONCEPTS.

159. Reasoning Defined.—Reasoning is an act by which we discover new truths whose general characteristics we already know. Thus, to take one of the simplest examples: if a boy finds a piece of wood, he knows, without trial of this particular stick, that it will burn. This knowledge is gained not from any experience of the individual fragment, but from his familiarity with the class to which it belongs. He knows that this bit of wood will burn because all wood is combustible. On like grounds, if you name to me any unknown person, I can predict his future death with certainty, because I know that all men are mortal. In other words, what is true of the class of which he is a member, is true of him as an individual of that class.

This act of detecting general facts in special instances which lie in the same class, is as natural, both to the child and to the adult, as that of perception or any of the other operations that follow it, like all the normal movements of thought, which we have explained, it takes place in the mind spontaneously on the presentation of its material; and the exclusive material on which reasoning operates, is classified or general knowledge.

160. All Reasoning Deductive or Inductive.—All reasoning proceeds either from a class, as a whole, to its members; or from its members to the class, according to the following axioms:

1. Whatever is true of the class as a whole is true of every member or sub-class that it contains.

2. Whatever is true of all the members of a class, as parts, is true of the class as a whole.

Whenever we infer, according to the first axiom that any characteristic which belongs to a class, belongs also to this or that individual or sub-class, contained under it, the reasoning we employ is deductive.

When, on the other hand, after finding that some fact is common to a sufficient number of individuals, we infer that it is a characteristic of the entire class to which they belong, we employ inductive reasoning.

161. Reasoning a Comparison of Three Concepts.—Scrutinizing the process of reasoning further, we find that it consists always in the comparison of at least three concepts, the first of which is the concept of a genus, the second is the concept of a species under the genus, and the third is the concept of a sub-species or of an individual which is contained in the species; as follows:

A bird is a biped;
A sparrow is a bird;
Therefore, A sparrow is a biped.

This comparison shows that the class, bird, is included in the class, biped, and, inasmuch as the class, sparrow, is included in the class, bird, it will also be included in the class, biped.

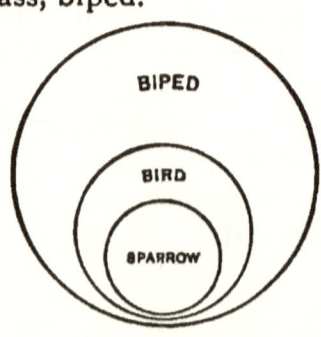

In other words, "what belongs to a class, belongs to all the members contained in that class." The class, bird, belongs to a wider class of animals that have two feet, and the sparrow belonging to the class, bird, has, consequently, two feet also.

> All fruit is perishable;
> The banana is a fruit;
> Therefore, The banana is perishable.

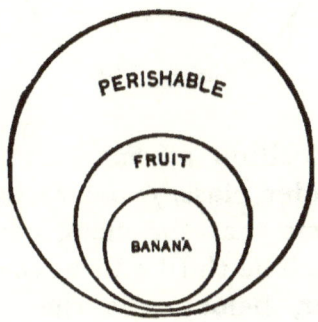

In this example, the class, fruit, is affirmed to be comprised in the wider class, perishable things, and the banana, being comprised in the class, fruit, is also comprised in the wider class, perishable things, which includes the class, fruit.

The formal statement, in the above examples, by which two concepts are thus compared with a third; and a conclusion deduced therefrom, consists of three judgments or propositions, of which the first affirms that a certain species, lower class, belongs to a certain genus, higher class; the second affirms that an individual, or sub-species, belongs to the species; and the third declares that the individual or sub-species, as belonging to the species, belongs also to the genius that includes that species.

Metals are fusible by heat;
Copper is a metal;
Consequently, Copper is fusible by heat.

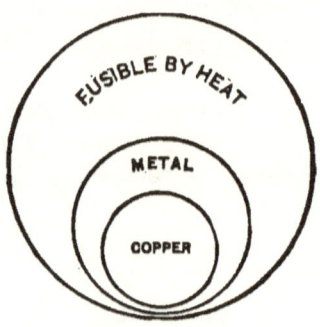

The first proposition affirms that the class, metal, belongs to the wider class, *fusible by heat.* The second proposition affirms that the class, copper, belongs to the class, metal. The third affirms that, consequently, the class, copper, belongs to the class, fusible by heat.

162. The Syllogism.—Three propositions so related to each other that the third is inferred from the other two, constitutes a syllogism. In constructing a syllogism, it is necessary to find a middle term or concept, which is so related to two other concepts, namely, a wider and a narrower one, that these may be compared through it and a conclusion drawn therefrom. Thus in the syllogism above, metals is the middle term by means of which copper and things fusible by heat are compared with each other.

163. Two Kinds of Deductive Reasoning.—There are two kinds of reasoning in deduction, namely; reasoning in extension and reasoning in comprehension. Reasoning is in extension, whenever the class denoted by the subject in each of the propositions constituting the syllogism, is contained in the class expressed by the predicate; and the reasoning is, consequently, towards the higher class.

In the syllogism:

> Man is mortal;
> Bismarck is a man;
> Bismarck is mortal.

The class man, which is the subject, is contained in the higher class mortal, as one of its members. The individual, Bismarck, subject, is affirmed to be one of the class, man, predicate, the reasoning, therefore, being towards the higher class, mortal, is in extension. It will be noticed that all the syllogisms given above are examples of reasoning in extension.

164. Deductive Reasoning in Comprehension.— The following syllogism is an instance of reasoning in comprehension.

> The United States include the State of New York;
> The State of New York includes Long Island;
> Therefore, The United States include Long Island.

The reasoning is in comprehension wherein we regard the predicate as a quality which the subject contains, rather than a glass which contains the subject. This we may do at pleasure.

> George is honest, i.e., honesty is one of his traits.
> Honesty includes truthfulness;
> George is, therefore, truthful.

In this example of reasoning in comprehension, the subject, George, contains the predicate honesty; the subject honesty contains the predicate, truthfulness; consequently, George possesses truthfulness as a characteristic.

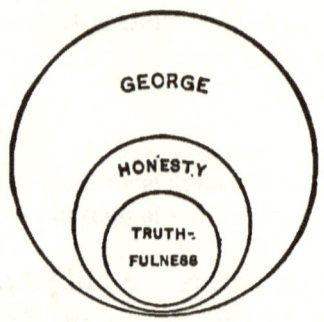

165. The General Truths from which Deductive Reasoning Proceeds.—We will conclude this elementary view of deductive reasoning, with a brief notice of important questions respecting the general truths or classified facts, from which it proceeds. Deductive reasoning is based on the axiom that what is true of a class or a whole, is true of all the members of that class or parts of that whole:

Now what warrant have we that the proposition which affirms something of the class or the whole from which we reason, is true? When we say:

> All men are mortal;
> Arabs are men;
> Then, Arabs are mortal.

How do we know that "all men are mortal." Every one accepts this proposition without question; what is the real basis of its truth? The answer is that its truth has been established by induction. In other words, the truth of a general proposition, "All men are mortal," has been verified by the knowledge that a vast number of individual men, including all in past generations, have died. This knowledge, which is the result of an induction of particulars, has been gained through experience, through history, and the testimony of others; and the truth of the general proposition has been so fully established thereby, that it is beyond question. There are, however, many general propositions which are based on an induction far less complete. For example:

> The early migration of birds is followed by a severe winter;
> The birds are migrating early this season;

Therefore, The coming winter will be severe.

166. Premise and Conclusion on the same level.—From the above considerations it is evident that the conclusion reached in any correct syllogism, will have only the degree of certainty that belongs to the wider proposition from which it is derived. It is as credible that a hard winter will succeed the early migration of birds in any given year, as that, in general, hard winters succeed the early migration of birds. The two propositions—special and general—stand on the same level in respect to verity.

167. Self-Evident Truths often used as the Basis of Deductive Reasoning.—Our conclusions, however, in deductive reasoning, are not all derived from general propositions established by previous induction. Not unfrequently, the general fact which serves as a basis of the syllogism, is an intuitive or self-evident truth. Especially is this the case, when the reasoning is mathematical. For instance:

Equals being added to equals, the sum are equals:

$A = B$ and $D = C$;
Therefore, $A + D\ 9 = B + C$.

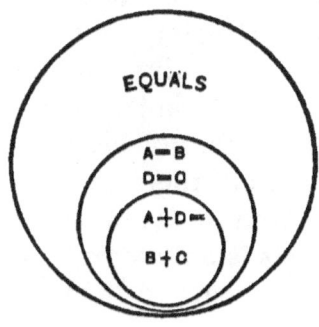

In cases like this the conclusion, like the premise, is absolutely certain.

168. Premises in Reasoning often Suppressed.—In ordinary reasoning the language employed is seldom arranged in the form of a syllogism. Often the conclusion and minor premise, are given while the the major premise or main proposition, is understood.

George cannot endure hardship, he is a mere child; if expanded, would be

>Children cannot endure hardship;
>George is a child;
>
>Therefore, George cannot endure hardship.

Walter suffers from intense cold because he is an invalid.
>Invalids suffer from intense cold;
>Walter is an invalid;
>
>Consequently, Walter suffers from intense cold.

169. Inductive Reasoning.—In reasoning from induction we reverse the deductive process. In deductive reasoning, we infer that what is true of a class or whole, is true of any of its members or parts.

In inductive reasoning, we infer that what is true of its members or parts, is true of the containing whole. In the proposition, "Heat expands all bodies," we have a general statement, based on the observation of a multitude of particular bodies under the effect of heat. Systematic experiments have been made on a great variety of bodies, with the uniform result of increased volume under increased temperature. Hence, an effect on its members is an effect on the entire class. The effect of heat is to expand all bodies constituting matter; therefore, heat expands matter.

Again: quinine is an antidote for malarial fever. This is an example of a general proposition which has been reached through a multitude of special instances, wherein quinine was successfully given as a remedy for this class of diseases.

It is clear that our knowledge of general facts and principles, is largely acquired through induction, and that observation and experience are the means by which such knowledge is primarily gained. In scientific induction, all the members of a class must be rigidly scrutinized, before the conclusion is reached that the quality under consideration is, or is not, its common characteristic.

170. Hasty Induction.—In the ordinary opinions of men, hasty or inadequate induction is the source of many prevalent errors. From noting carelessly the peculiarity of a few individuals, the superficial thinker rushes to the conviction, that such peculiarity is the uniform trait of a wide class. For example, several business men of Frankfort-on-the-Main, having been overreached in two or three commercial transactions with New York, leaped to the conclusion, that "America is a nation of swindlers." Even a distinguished English scientist, once traveling in the West, noticed some informality at a hotel table, and derived therefrom the settled opinions, that "Western people are crude and clownish in table manners." Western farmers have suffered in a few cases by sending their products to a distant market and have adopted therefrom the opinion, that, "Middle men are rapacious and dishonest."

INSERTIONS ON CHAPTER XII.

Define and illustrate the act of reason. Reasoning primarily is a spontaneous operation. What are the two kinds of reasoning? Upon what axiom is deductive reasoning based? What axiom underlies inductive reasoning? When is reasoning deductive and when inductive? What number of concepts is indispensable in reasoning? What is the relation of these concepts to each other? How is it proved that any sparrow is a biped? Show the relation of the three concepts employed by circles within circles. Prove that the banana is perishable, and show the process by circles. Give the formal statement in the above examples by which two concepts are compared with a third, and a conclusion deduced therefrom. Show and illustrate by circles the process which proves that copper is fusible by heat. Construct a syllogism and explain its parts. What are the two kinds of deductive reasoning? Give an example of a syllogism which illustrates reasoning in extension. Construct a syllogism which illustrates reasoning in comprehension. Show by circles the difference between these two kinds of deductive reasoning. Upon what truths is deductive reasoning based? What warrant have we of the truth of a general proposition? The premise and the conclusion on the same level of certainty? Self-evident truths as the major premises in deductive reasoning. The major premise often understood in ordinary reasoning. Give examples. Inductive reasoning reverses the process of deduction. Give examples. Our knowledge of general facts and principles acquired mainly through induction. What is the character of hasty induction? Hasty induction the source of many errors. Give examples.

Chapter XIII.

THE INVARIABLE SERIES OF MENTAL ACTS THAT IN THE GROWTH OF MIND BEGIN WITH THE SENSES AND END IN REASONING.

171. The Tri-unity of Power, Act and Product.—In the preceding analysis, which reveals a fixed order in the succession of primary intellectual action, we have not always paused formally to distinguish the faculty from its action, nor its action from the resulting product. A careful discrimination of these three from each other is essential to a complete knowledge of the facts involved. And first a faculty is the power which the mind has of acting in a specific direction. What is called the action of a faculty, is really the mind acting in this specific direction. So when we speak of the product of a faculty, as a concept or idea, we mean simply the condition of mind which any special act produces when completed. A faculty, its action and its product are not, then, three entities existing in the mind and separate from it, but rather three conditions or modes of the mind itself.

172. Distinction Between the Object of a Faculty and its Product.—A further distinction should, in this connection, be carefully noted. The object on which a faculty expends its effort and the product that results from that effort, are not identical. The concept presented by a preceding power, is the object or material which the action of a faculty transforms into its own peculiar product. Thus the product of a faculty constitutes the object of its immediate successor. In this way, every object is, by the modifying process it undergoes, changed to a product of a different character And thus a series of mental transformations begins with the senses and ends with reasoning.

Let us follow this uniform consecution of object and product a little more specifically. When the eye falls on a visible object, the product is the instantaneous effect on the mind, which we term a precept. Now this precept, the product of sight, is the object which memory instantly appropriates and transforms thereby, into an unconscious, concrete concept. The mind recalling and presenting to the faculty of conception this unconscious concept, which is the product of memory, it becomes thereby the object of the conceiving faculty, which, by vivid representation, changes it from an unconscious concrete concept to a conscious one. Then this conscious concrete product of conception becomes the object out of which analysis evolves as its products, the simple concepts of properties and parts; and each of these again is the object from which, by repeated comparisons, abstraction complete its products, namely, the abstract concept. Next taking as its object the concrete concept, the simple concepts of its properties, and the abstract concepts derived from these, the classifying faculty develops therefrom its own peculiar product. The class concept, which in turn assumed as the object of judgment, is wrought by *its* affirming power into a predicated concept of a class characteristic. Lastly, the reasoning power, accepting the products of judgment and reducing them to the peculiar combination displayed in the syllogism, finally elaborates therefrom the product appropriately called an inferred concept.

173. Tabular View of the Succession of Powers, Objects, Acts and Products.—These facts when fully comprehended, prepare us for a tabular view which shall present the intellectual faculties, their objects, actions and products, in the order of their birth, growth and development.

Faculty.	Object.	Action.	Product.
1. Sense-perception.	External. Object.	Touching. Seeing. Hearing.	Percept.
2. Memory.	Percept.	Receiving. Retaining. Recalling.	Unconscious. Concrete. Concept.
3. Conception.	Unconscious. Concrete. Concept.	Representing.	Conscious. Concrete. Concept.
4. Analysis.	Conscious. Concrete. Concept.	Analyzing.	Simple concepts of properties and parts.
5. Abstraction.	Simple concepts of properties and parts.	Abstracting. Classifying. Affirming.	Abstract concepts. Class concepts. Predicated concepts of class characteristics.
6. Classification.	Concrete concepts. Simple concepts. Abstract concepts.		
7. Judgment.	Class concepts.		
8. Reasoning.	Predicated concepts of class characteristics.	Reasoning.	Inferred Concepts.
9. Imagination.	Concrete concepts. Simple concepts. Abstract concepts.	Imagining.	Concrete. Image. Concepts.

174. In What Our Knowledge Consists.—It is manifest that the knowledge which the mind gathers, as the years pass, will consist of the various concepts named in the series. It is also manifest that, in the juvenile period, these concepts which represent knowledge in its different forms, acquired successively as indicated, will go through the serial transformations with comparative regularity and completeness under favoring conditions and judicious guidance; but, lacking these, they will be the imperfect products of efforts that are irregular and fitful. Hence the imperative necessity of systematic primary instruction that shall accord with the inevitable successive operations which the mind initiates in the early unfolding of its powers. In the elementary school before Pestalozzi's time, nature and the teacher were antagonistic forces; the first began at one end of the series; the second, at the other.

175. Maturer Faculties not Confined to the Order of Growth.—It may suffice to note here that the maturer action of a faculty is by no means limited to the class of objects which the unfolding processes present to it. Thus while the faculty of conception is, in the incipient stage, limited to the act of representing concrete concepts, it also represents or holds up to consciousness, all the subsequent concepts as soon as they are formed.

It is enough to say at this juncture, that the first office of conception in the order of time and in the unfolding activities, is to represent the concrete concepts transmitted by the senses.

Premising these important considerations, I present, as leading to a more complete familiarity with the facts developed in the foregoing pages, a tabular statement of the various concepts in the order of their production and apart from the names of the faculties which produce them.

1. Sense percepts.
2. Unconscious concrete concepts.
6. Image concepts.
   { 3. Conscious concrete concepts.
     4. Simple concepts.
     5. General abstract concepts. }
   6. Class concepts.
   7. Predicated concepts.
   8. Inferred concepts.

176. The Successive Intellectual Acts when the Class is a Familiar One.—Having learned by a careful inspection, that every new external object to which the mind attends, is modified by successive operations, let us gain the highest familiarity possible with these operations by specific examples. An appeal to personal experience reveals the subtle rapidity with which the mind flashes through the series of acts from sense to reason, whenever a single concrete idea is gained through the senses and referred to a well known class. It must be kept in mind that any description of the serial acts that follow a particular sense-perception, while it presents their nature and order, will fail inevitably to depict their swiftness. Thus a pedestrian catches sight in the distance of an individual object which he has never seen before. He perceives that it is a dog, a black and tan terrier, and infers its proclivity for catching rats, all in a single instant. Here the mind rushes through the series of consecutive acts with such celerity, that to ordinary consciousness, they seem fused into one. But any complete analysis does not fail to prove that this operation, though seemingly single, is really a logical series from which no one of the serial steps is absent.

As the first step, the mind notes the single impression made on the eye, thus cognizing the object as a whole; this may be called an individual percept.

Secondly; memory carries forward this percept as a whole from one indivisible instant to another and thus makes it possible for the mind to contemplate it. It is this subtle connecting together of the percept from the immediate past with that of the present, that enables the memory to gather and hold a concrete concept.

Thirdly; the mind aided by the presentation of both sense and memory, is enabled to gain and represent to itself a concrete concept which is all the more complete and full because of its perpetual renewal by the continued presence of the external object; this is a conscious concrete concept.

Fourthly; come the subtle flashes of analysis which note the outlines, form, motion, size, color, parts of the concrete concept as a whole; these are simple concepts of the properties of an individual whole.

The fifth step is the subtle spontaneous reference of these properties to the corresponding abstract concepts previously gained. This step makes the comparison of the characteristics of this object with those of like objects, possible. It is by the light which previous abstraction affords that we are able to recognize the likeness or unlikeness of characteristics on which classification is founded.

The sixth step consists in the adjustment of the object observed in its appropriate class or classes. This adjustment is based on the comparison of its properties with the constant properties of the class to which it is found to belong. The object which the pedestrian noticed is found by analysis to have life and motion; it is an animal. Analysis reveals further certain characteristics of organism and outlines; it is a dog. It shows also additional peculiarities of size, shape and arrangement of colors; it is a black and tan terrier.

But seventhly; judgment affirms that all black and tan terriers are rat catchers, that this dog is a black and tan terrier; and reason crowns the series by inferring that, therefore, the terrier just classed as a black and tan, is a rat catcher.

Thus we find that what seems to the careless observer a single mental act completed while the eye rests on an outside object, is a serial number of acts commencing in sense-perception and culminating in reason. The object unknown as a concrete individual, is perceived as a unit; held in memory; represented to the mind; analyzed; compared with like objects; classified; subjected to class judgments, and through these, modified by the reasoning faculty. No appreciable intervals of time separate these acts; they are closely welded together under a single mental impulse yet their number and order are, under like conditions, always the same.

177. The Succession of Intellectual Acts when Exerted on Objects that are Unfamiliar.—It is evident that when both the concrete object under scrutiny and the specific class to which it belongs, are unfamiliar or unknown, each of the series of mental acts exerted upon it, becomes more protracted and deliberate and is thereby consciously separated from the act that follows it.

The scientist finding in a remote region an unknown bird, for example, takes each of the intellectual steps antecedent to classification, with scrupulous care. All his faculties are on the alert for complete and exact examination. Consequently, the initial effort of perception is exhaustive; the memory following distinct and tenacious; the concepts vivid and full and protracted; the analyses minute and deliberate, noting with precision all the internal and external organs together with their constant characteristics of size, color, shape, habit and habitat, and the subsequent comparisons of these characteristics with similar or dissimilar ones noted in previous analyses of the bird, are made repeatedly and with nicety. In the resulting act of classification, the species, it may be, is discovered to be a new one and the classifying act consists in adjusting the newly found species under a family and genus already familiar. Then follows judgment affirming that the family of which this new species is a member, possesses certain characteristics which the analysis above did not reveal. If we suppose, for instance, that the species lately discovered belongs to the thrush family, the fact which judgment predicates, is, let us say, that all members of the thrush family are singers and that the species of bird under scrutiny is a member of the thrush family; whereupon reason concludes that this new species must be musical also. The above example may serve to show that every effort of earnest investigation tends to expand the series of intellectual acts, making each consecutive operation prolonged and distinct, and giving it a conscious individuality.

QUESTIONS ON CHAPTER XIII.

What is the distinction between a faculty, its action, and its product? Also, between the object of a faculty and its product? How is an object changed by the action of each faculty into its product? In what manner does the succession of the object, action, and product of each faculty constitute a series which begins with sense-perception and ends with reasoning? Write out the table showing the succession of powers, objects, acts, and products? In what does the knowledge of a mature mind consist? Mature action not limited to the objects presented by the unfolding series. Give a tabular statement of the various concepts in the order of their production. What is the character of the successive acts when the object perceived belongs to a familiar class? Explain and illustrate the rapidity of the six consecutive steps. Contrast the slowness of the succession of our intellectual acts when exerted on objects that are unfamiliar. Illustrate.

Chapter XIV.

INTUITION.

178. "Intuitive Ideas," Comprehending Concepts.—In treating of the foregoing series of mental operations, we noted two faculties whose sole province is to gather original knowledge. One of these is external perception, and the other internal perception. The first, called also sense perception, has the senses for its organs, and the phenomena of the material world for its objects. It is, as we have seen, the faculty with which mental activity begins. The second, viz., internal perception or concentrated consciousness, takes cognizance solely of the phenomena of the mind, consisting of its present ideas, feelings, and volitions. Now each of these two originating faculties, being limited in its action to its own field, gains therefrom concepts which have their distinguishing characteristics.

It is the characteristics of a concept that, as we have already seen, reveal its origin in the mind. If, for instance, analysis notes in any concept, the qualities that belong to matter, such concept was gained by external perception through the senses. Even when the imagination has modified a sense concept, or recast it wholly into a new form, the simples of which it is composed inevitably disclose their external origin. Thus pronounce the word "mountain," and whether I realized its meaning through imagination or through the recollection of a special instance, the characteristics of size, shape, color, etc., attest the fact that its original source was the senses.

In like manner we are able to determine the acquisitions of internal perception by scrutinizing their peculiar characteristics. When I recall from memory the concept of a beautiful landscape, my consciousness perceives this effort as a purely mental act wholly unlike any operation in perceiving an external object. This concept from memory is the immediate product of the mind, and is discriminated as such by internal perception. The feelings, likewise, as joy, grief, fear, love, hate, are simple conditions which the mind originates, and as such are exclusively the objects of internal perception.

179. Concepts Not Gained Through External or Internal Perception.—Now if we examine exhaustively any concept previously acquired through either external or internal perception, we shall find it invariably attended by other concepts that are totally different from itself both in character and in origin. In fact, any complete inventory of our conscious ideas will reveal the fact that every concept which the senses or consciousness has gathered, is dominated by a group of concepts which are wholly distinct from itself in their nature and kind. Such concepts, since they have no characteristics in common with those which external and internal perception have supplied, must have gained access to the mind through another channel.

Recalling and fixing the eye of consciousness upon the Washington Monument, for example, I instantly perceive from its qualities of size, shape, color, etc., that it is a product of sense perception. My conception now presents it as the distinct notion of a single external reality which sometime in the past presented itself to my senses. Scrutinizing the present concept of the Washington Monument do I find it a solitary one? Does my consciousness represent it as single and alone? Can I, by the most strenuous intellectual effort, separate and consider it apart from certain other concepts which are its close and constant attendants? Let us see?

180. The Concept of Space.—I cannot conceive of the Washington Monument without conceiving that it is *somewhere;* in other words that it occupies *space.* The concept of space encloses that of the Monument, and is inseparable from it. Recall the notion of the Monument, and you recall invariably a notion of the space that contains it; the one is an invariable accompaniment of the other. The same is true of all the concepts of external objects gained through the senses. Cities, islands, mountains, clouds, rivers, animals, and the whole countless host of material things, are perceived and conceived as somewhere, that is, as located in space.

Body and space, both as external facts, and as conconscious concepts, are perpetual concomitants.

If now we fix our attention on the concept of space, we perceive at once that it has none of the qualities that distinguish matter; it is without taste, odor, form, color, or resistance. We cannot smell, taste, see, hear, or handle space. It is, in fact, wholly imperceptible to the senses. As a concept it is not concrete, but simple, and it cannot therefore be analyzed. Having no qualities or characteristics it cannot be classified. Whether as an external reality, or conscious idea, it stands alone by itself as the infinite extension which contains all existing forms of matter, from atoms to planets. Inspecting more formally the concomitant concepts, body and space, the following contrasts are manifest:

1. Body is limited in extent, giving rise to figure, size, mobility, locality, etc. Space is unlimited in extent. It is impossible to conceive of space as bounded.

If, in imagination, we assign to space a boundary, our thought at once transcends it—our conception reaches out far beyond the material universe and finds no barriers. Body is finite, space is infinite.

2. Body is complex. We have found by analysis that every form of matter is composed of properties and parts. It is impossible to conceive of a body except as consisting of various characteristics which serve as the basis of its classification. Space, on the other hand, as noted above, is simple; it has no contrasting parts or positive qualities, and, consequently, belongs to no *class of ideas*. Body is concrete, complex; space is simple, unique, fundamental in thought.

3. The concepts of body are individual; that of space universal. All men gain and retain concepts of external things or bodies; but not necessarily of the same bodies; that is, the concepts are individual and different.

But the identical, unvarying concept of space is present to the consciousness of every human being.

The permanent presence of the concept of space is one of the fundamental conditions which make thought possible. Man would not be man without it. The concept of space is *universal*.

4. A concept of body is *contingent*, the concept of space *necessary*. A concept is called contingent when we can conceive of its object as *not existing*. Centering my consciousness upon the concept of the Washington Monument, I am at once aware that I am able to conceive of the non-existence of the Monument itself. I can think of it as utterly annihilated, leaving only empty space where it once stood. So of all things that occupy space—trees, forests, buildings, cities, suns, and systems; these may instantly be destroyed in thought as time finally destroys them in fact. The concept of

body in whatever form is contingent, but when dwelling mentally on the concept of that illimitable expanse which environs all material forms, we attempt to think of it as not existing, we are instantly conscious of an utter failure. Space presents itself to our consciousness as the idea of a reality which is indestructible, immutable, boundless, eternal. Not one jot or tittle can we add to, or subtract from, the infinite sum-total which conception holds and represents to our minds. It is an idea which the will can neither expel nor modify, nor can its objects be thought of as having no reality. The concept of space is necessary.

181. The Concept of Time.—But space is not the only one of the kindred ideas that uniformly cluster around every idea of body which the mind contains. Just as I am compelled by a law of thought to regard the Washington Monument as existing in space, so I am forced by the same law to recognize it as existing in time. There was a period when it was being erected, then followed a period when it was completed. It will probably endure through centuries, but there will come a date, as the years lapse, when it will yield to the inflexible law of decay. In other words, the Washington Monument occupies time, in another aspect indeed, but quite as undeniably as it occupies space. In fact, it is manifestly impossible to conceive of the structure in question, without conceiving more or less vividly of the time in which it began, endures, and will end. The concepts of body suggest inevitably the concept of time. The same contrast of characteristics, then, will be apparent on inspection between the concepts of body and time as between the concepts of body and space.

1. Body is limited in duration, has a beginning and an end; time has neither beginning nor end. It is not

in our power to think of time as commencing in the past or as closing in the future. The concept of body is *finite*. The concept of time, which forever attends it, is *infinite*.

2. Body, as we have seen, is complex, being composed invariably of unlike properties and parts; time is simple, having absolutely no characteristics that are distinct from each other. Its divisions, as marked by events, differing only in duration, are the inseparable parts of that absolute whole to which there are no conceivable limits.

3. Further, the concept of time, like that of space, is universal. It is, in other words, a conscious idea in every human mind, whether civilized or savage. To conceive of individual things, or a succession of events, without its presence in consciousness, would be impossible. Expunge the concept of time from the mind, and the power to think is annihilated. It is essential to every intellectual act.

4. The concept of time is likewise necessary. Strive as we may to conceive of it as not existing, the effort is utterly futile. We are able, as shown above, to picture in thought the negation of particular concrete things or events; but the infinite duration which serves, so to speak, as their uniform background, has no thinkable negative.

182. The Concept of Cause.—Another concept belonging to the group that invariably accompanies that of body is the concept of cause. It is not in our power to think of any object on which the senses may dwell except as an effect. The human mind is compelled by the law that prescribes its mode of action, to conceive of every material thing as the product of a power which is distinct from itself. The pen I hold, the paper on which I write, the painting that adorns the wall, the shrubs and flowers on the border outside, were not self-created. Thinking of each one of these, I am conscious also of a concept of some producing cause or causes of which it is the effect. I may not know, indeed, precisely what these causes are, or how they operate in effecting the changes that precede the result. What I do know is that the several articles named are not self-produced, but, on the contrary, are the offspring of a power or powers distinct from themselves. In short, it is not in the nature of thought to conceive of a concrete object or event as *uncaused*. The concept of body, or any of its modifications, is invariably attended by the concept that it has a cause. Reflecting on the concept of cause, or force, as contrasted with that of body, we find that, like those of space and time.

1. It is simple : in other words, the concept of cause or force cannot be analyzed into different parts like that of a complex object. The concept of cause consists in the ever-present conviction that every object under scrutiny is the product of a distinct creative energy. The concept of cause cannot be separated into different elements, and is therefore simple.

2. The concept of cause is likewise universal. It is one of the essentials to the process of thinking. No human being can conceive of a thing, or a change, without the concomitant concept of a cause for it. This concept or cause may indeed be dim and vague in the consciousness of a savage or a child, but it is nevertheless invariably coupled with each one of his concrete concepts.

3. The concept of cause is necessary. Given the concept of any finite thing, and we are manifestly unable to conceive it as *uncaused*. The mind affirms, absolutely, that the object of every idea acquired through

the senses is an effect. The concepts of cause and effect are uniformly coexistent, inseparable. The thing caused is contingent, i.e., may be conceived as not existing; but *as existing*, the non-existence of its cause is inconceivable. The concept of cause is necessary.

183. Substance.—There is still another concept of a character akin to that of time, space, and cause, which centres perpetually in each of the concrete concepts; it is the concept of *substance*. If we analyze a concept of any material object, say a marble slab, we inevitably find that its component elements, as gathered by sight and touch, are properties simply. The eye takes cognizance of only its length, breadth, thickness, shape, size; the hand perceives its resistance, smoothness, and temperature solely. These qualities of the piece of marble are all that the senses can discern, but these qualities do not constitute the marble slab. Qualities are only the manifestations, the phenomena of that in which they cohere. Extension, resistance, shape, size, are the qualities of something that extends, resists, has shape and size; something which is not obvious to the senses: that something is substance. The substance, then, is the *reality* of a body, and the qualities are its manifestations. Perceiving the qualities of a body through the senses we are compelled in the nature of thought to *infer* the substance. Thus the group of qualities which each individual object presents, and the substance that contains them, are forever associated in thought. Given the concept of the group of qualities that sense perception has gathered, and the concept of the substance is inevitably present to consciousness. The nature of this concept of substance is kindred in important particulars with that of space, time, and cause.

1. Substance is simple; with the utmost exertion of thought we cannot analyze it. The properties of any given body are many and various, but the substance which includes them is a unit. We do not know what this unit is, but we do know its existence.

2. The concept of substance which centres in every concept of body, as a group of properties, is present with more or less vividness in every human consciousness. Property and substance, though opposites in character, are inseparably connected in thought. Clearly, the concept of substance is one of the group of ideas that make thinking possible. The concept of substance is universal.

3. Any attempt to realize the concept of the properties of a body as revealed to the senses, without realizing also the concept of the substance in which they reside is unavailing. In fact, having the notion of a group of qualities, the mind is wholly unable to conceive of the absence of the substance they qualify. The concept of substance is necessary.

184. The Conscious Operations of Mind Necessitate the Conceptions of Time, Cause, and Substance.—Turning now to the realm of mind we discover with equal certainty that every thought, feeling, and volition is invariably attended by the necessary concepts under consideration, except that of space. *The concept of space* is manifestly absent from our mental movements as such. We are not able to harmonize the notion of extension with any modification of mind. It is inconceivable that a feeling, for example, has length, breadth, thickness, figure or size. Love is not a cube or a sphere. Ideas, emotions, judgments, cannot be construed to thought as occupying space.

185. Time Suggested by the Manifestations of Mind.—Consciousness presents a constant succession of mental operations in their perpetual round. Thought, feeling, choice, constitute the successive links in the chain of experience. It is the sequence of events, whether as revealed to external or internal perception, that suggests inevitably the concept of time.

186. Cause likewise Associated with all the Operations of Mind.—It is also undeniable that every product of mind, whether of the intellect, the sensibility, or the will, must have a cause. One can no more conceive of an uncaused change in his mental state, than of an uncaused event in the other world. The concept of cause clings as closely and as constantly to all the mental products as to the material ones.

187. Mental Operations Suggest the Concept of Substance.—Mind in its essence is beyond the reach of consciousness. We are conscious only of its manifestations. When the faculties, feelings, and will are wholly dormant, internal perception has no objects and the mind no consciousness of its own existence. This is our mental condition in a swoon, in catalepsy, and in perfect sleep. The mind in its waking hours perceives immediately the presence of thought, feeling, volition, and is compelled to infer the existence of that which thinks, feels, and wills. Loving, perceiving, choosing, are conscious phenomena; but these acts must each have an actor that puts it forth, and this actor is mind or substance. Since the phenomena of mind and those of matter are of a totally different character, we may justly infer that a material and a spiritual substance are correspondingly different entities.

188. Intuition.—The power which, without any effort of will, presents to the mind the necessary concepts we have considered above, may be called *intu-*

ition. Thus the mind has three originating faculties; viz., external perception, which discerns directly the qualities or phenomena of matter; internal perception, which discerns the operations or phenomena of mind; and intuition, which perceives and verifies the great realities that underlie the phenomena revealed to external and internal perception.

189. Intuition—Its Early Activity.—The first movements of intuition are awakened by the earliest conscious perceptions of the intellectual senses. It is inconceivable that the infant should perceive resistance through touch without perceiving that there is something which resists. The incipient discernment of external colors necessarily includes a discernment of the object colored, i.e., of substance. In gaining its first notion of the extension, shape, or size of a body, the child cannot avoid the notion of its place or position, which, when expanded, is the notion of space. So soon as he discovers that one incident follows another, whether in things around him or in his own thoughts, there dawns upon his mind, however obscurely, the notion of time. So soon as he observes actual instances of causation; when, for example, he knows that his own will moves his hand, foot or tongue; that his mother caresses and cares for him; then, the dim idea of cause is awakened in his mind.

These concepts of space, time, cause, and substance, vague and rudimentary at first, gradually grow in distinctness as the experiences that give them birth, become extended and clear, until consciousness recognizes their presence as essential to every mental operation. From first to last they are spontaneities. The will cannot originate, modify or expel them from consciousness. No human intellect is capable of imagining their absence either from thought or from being. They are *race* ideas; innate, involuntary, and as indestructible as mind itself.

190. Terms which Designate the Objects of the Originating Faculties.—Any single object which is cognized by external or internal perception, is, in philosophical language, termed a Phenomenon. The collection of qualities which a granite boulder presents to the senses, are its "phenomena." Any single mental manifestation which consciousness discerns is called a "Phenomenon" of mind. All intellectual acts, as perceiving, analyzing, imagining, etc; all the actual feelings we experience and every volition, of which internal perception takes account, are the "phenomena" of mind. External perception through the senses cognizes exclusively the phenomena of matter. Internal perception through consciousness cognizes the phenomena of mind. Each of these two originating faculties is limited in its operation to the class of phenomena which constitute its peculiar objects: the one to material, the other to mental phenomena.

On the other hand, that simple entity to which any one collection of phenomena belongs, is designated by the word "noumenon." In the granite boulder that of which the phenomena of extension, size, figure, hardness, are simply manifestations and which intuition affirms must be present, is its "noumenon." The properties of matter which reveal its presence to the senses, are, I repeat, phenomena of matter. But the actual things in which these properties cohere, as properties, are the *noumena of matter*. The noumenon is the *reality;* its phenomena are the *manifestations* of that reality.

In like manner the conscious operations of an individual mind which are termed its phenomena, are simply the manifestations of the substance or *noumenon* which constitutes the mind itself, considered apart from its operations. That person or being who is designated by the noun "self" or the pronoun "I" is the noumenon of which all I think, or feel, or purpose, are the phenomena. In general, the substances which underlie or present phenomena of matter, may be called the noumena of matter, while those which present mental phenomena may be named the noumena of mind or spirit.

Thus we present in tabular statement

Originating Faculties.	Objects.
External Perception	The Phenomena of Matter.
Internal Perception	The Phenomena of Mind.
Intuition	The Noumena of Matter and Mind.

191. Concepts of Phenomena, the Chronological Antecedent; those of Noumena, the Chronologial Consequent.—We will now consider the contingent and the necessary concepts as to the order of their acquisition. Though the concepts of phenomena and of their corresponding noumena, are indissolubly connected in the mind, the intellectual acts by which they were acquired, were not absolutely simultaneous. The concepts of phenomena uniformly precede the concepts of noumena in the order of time, though it be but an inappreciable instant. In other words, external and internal perception must discriminate their objects before intuition can perceive the fundamental entities which make the existence of such objects possible. It is an inflexible law of perception that contingent knowledge should *introduce* the necessary knowledge that underlies it. Thus it is manifestly not in our power to gain a notion of a special material substance, without first cognizing the qualities it displays to the senses. I am compelled, for instance, to notice the characteristic properties of a piece of glass, before I can determine *that it is glass;* for glass is the name that designates a special substance, and not the assemblage of properties that characterize it.

It is equally evident that the mind must perceive, through the senses, the length, breadth, thickness and shape of a contingent body, before it can perceive, through intuition, the space without such a body, having no location, could not exist. In other words, our knowledge of limited extension in matter, precedes our knowledge of unlimited extension in space. In like manner, the perception of an interval between two successive events or experience, is the only condition on which the mind can gain an idea of time without which events could not occur. Here, also, limited time, as measured by two events, is the invariable antecedent in the order of acquisition of time unlimited, past or future. The same order clearly exists in the acquisition of the ideas of effect and cause, whether special or general. Before I can gain a notion of the cause of a body or of a particular event, I must notice or think of it as *an effect.* I must perceive a wound, for example, before I can think of the blow that produced it. It is by the generalizing of particular events as effects, that come under our observation, that we attain the necessary truth that " Every event must have a cause." We have thus reached a law or axiom in psychology, viz., that the concepts acquired by external or internal perception, precede, in the order of time, the concepts acquired by intuition. Or, more concisely expressed, the contingent ideas are the chronological antecedents of the necessary ideas.

In scrutinizing our necessary concepts further, another fundamental fact or law becomes apparent.

After having gained the concepts of phenomena and their corresponding noumena, I am unable to conceive of the existence of the former without the latter. I cannot, for an instant, think of a succession of events, without conceiving of the pre-existence of time in which they occur. I find it impossible to conceive of a change, except as the product of an antecedent cause. I fail even to think of properties, except as the mere manifestations of a substance that gives them reality. I have no notion of body without the unavoidable conviction, not only that it could not exist, except in space, but that space is the antecedent in the order of existence.

This while the concepts of phenomena are unquestionably antecedent in the order in which the mind acquires them, the concepts of the noumena when once acquired, involving, as they do, the inevitable conviction that their objects are necessary to the existence of the former, are consequently antecedent in the order of being. If it is impossible to think of body without space; event without time; change without cause; property without substance; then space, time, cause, and substance are not only essential to the existence of these phenomena, but as concepts they are essential to thought itself. Since the ideas of body, event, and property precede ideas of space, time, cause, and substance in the order of acquisition, the former are the chronological antecedents and the latter are the chronological consequents. But since the ideas of space, time, cause, and substance are indispensable conditions to our conception of the existence of body, event, property, they are fittingly termed their logical antecedents, while the ideas of the latter are termed their logical consequents.

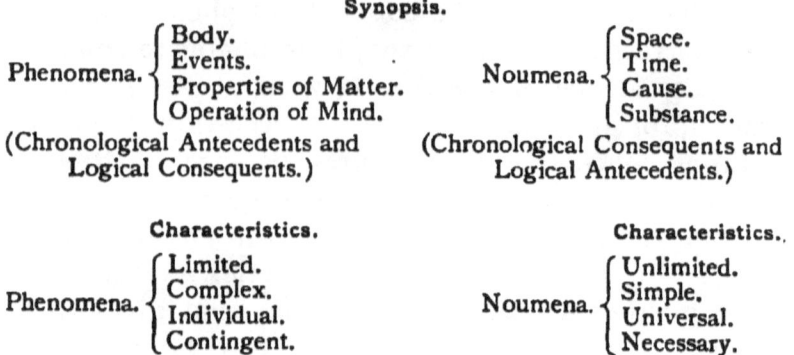

192. Further Facts respecting our Necessary Ideas.

—It was shown in a previous chapter that thought, when running in well-worn grooves, completes a series of distinct operations in a flash. So instantly does the mind blend the successive steps of a familiar process into one, that consciousness cannot distinguish them from each other. This is especially true of all our spontaneous operations. Their result is clear to the mind in every instance, but the different movements by which it was reached are often so subtle as to elude our notice. The same fact is apparent in the succession of contingent and necessary ideas as to the order of their perception. We necessarily discern the qualities of any individual body before we gain the idea of its substance. The concepts of body precede those of substance, space, time, cause, but precede them only by an unconscious, indivisible instant.

We cannot fathom the mystery of intuition or explain its mode of action, but since its products, which are utterly different from those of internal and external perception, are manifest to every human mind, the inference that such a faculty exists is inevitable. Its existence as a fundamental power in the human soul, evinced by the conscious presence of the ideas it generates, is a rigid deduction from the necessary truth that every effect must have a cause.

Beyond question, the faculty of intuition is rudimentary and limited in the infant and the savage. Its vigor and effectiveness are undoubtedly increased by culture. The clearness of its products depends on the completeness of the contingent concepts which, as we have seen, are its chronological antecedents. But their final distinctness as permanent ideas, is to be reached only by making the necessary concepts special objects of study and reflection.

Manifestly the faculty of intuition is one of the constituent powers of the human mind and its strength in any individual intellect is, like that of all the original faculties, derived from inheritance and the discipline resulting from exercise on its peculiar objects. It is therefore susceptible of the highest culture. Since the necessary concepts which intuition originates, underlie and make possible every process of thought, its activity begins with infancy and ends with life. For this reason the education which it receives must go along with that of every faculty in the intellectual series. Certain studies, however, as will be seen hereafter, furnish to intuition severer and more exclusive exercise than others, and these are, consequently, more available in special periods of its growth.

193. Our Knowledge of Necessary Truths.—It may be pertinently asked what is the character of our knowledge of those forms of existence called necessary truths? In other words, what does the mind positively know of the noumenal entities, space, time, cause, substance and the like? The answer which follows the scrutiny of our ideas of these entities, is obvious. We know nothing of the nature of space, time, cause, substance, etc., but we do know their existence with absolute certainty. I cannot comprehend space, or know what it is; but I can apprehend space, or know that it exists. Indeed, no knowledge that I possess, is more complete and undeniable. I perceive events and understand their character, but *time*, which is essential to their occurrence, is revealed to me as an unquestionable reality without characteristics and, therefore, incomprehensible. Its positive existence, however, I am utterly unable to doubt. It is not in my power to question the fact that every event, whether external or internal, is the product of a cause, but of the nature of cause or force I am hopelessly ignorant. When conscious of thoughts, feelings or motives, I can discriminate one from the other; can perceive the peculiarities of each and assign it to its class; but the spiritual substance which gives them birth is an unfathomable mystery in all, save its bare existence. When the concept of any material form is acquired, I know absolutely, that space is *that which contains it and is itself boundless.*

If I observe one event following another, I have the conviction that time is that in which they occur and that in itself, it is endless; this conviction is instantaneous and inevitable.

Given the conscious knowledge of a contingent fact, material or mental, and I affirm, at once, with absolute conviction that its existence must be due to a cause: an antecedent power which produced it. Considering this cause attentively, I find that I am compelled by a law of thought to accept it also as the effect of a preceding cause, which in turn, is the product of another cause. A step still higher in a series that ends finally in the *great first cause, or God Himself.*

QUESTIONS ON CHAPTER XIV.

What two faculties have we noted whose province is to gather original knowledge? With which of these does mental activity begin? What concept does each of these two originating faculties gather? By what means can we determine the origin of a concept in the mind. Give examples. What concepts does the mind contain which do not originate either from external or internal perception? Give examples. In conceiving of the Washington Monument what other concept is inseparable from the idea I gained? What concept not gained through external nor internal perception attends all concepts of external objects? Describe the concept of space, and distinguish it from the concepts of matter. Give the four characteristics of space which are in contrast with the characteristics of body. What other necessary concept inevitably accompanies the concept of the Washington Monument? Give the characteristics of the concept of time. Give the four particulars in which the concept of time is contrasted with the concept of body. What third concept is also indispensable to the concept of body? The impossibility of conceiving of body except as an effect. Give also the three particulars in which the concept of cause stands in contrast to the concept of body. What is a fourth concept which is perpetually present to every concrete concept? Explain the nature of the concept of substance, and give its characteristics. Give three characteristics in which the concept of substance is in contrast with the concept of body. Which of these four concepts is absent from our mental movements as such? Succession of mental operations suggests the concept of time. Every product of mind suggests a cause. Every manifestation of mind suggests the idea of substance. Power to originate necessary concepts called intuition. Origin of the necessary ideas and their early manifestation in the mind. Vagueness of these ideas in early life. Define the word phenomenon, and explain its use. Phenomena the products of two faculties, viz., external and internal perception. Propriety of the use of the word noumenon and its plural noumena. Give the relation of noumena as perceived by intuition to phenomena as perceived by external and internal perception. Give the objects of each of three originating faculties as shown in the tabular statement. What is the order of time in which we require the contingent and necessary concepts? What is the antecedent mental action that give us the notion of space? What perception precedes concept of time? What contingent idea inevitably precedes the notion of cause? What are the chronological antecedents of our necessary ideas? What is the chronological antecedent of the concept of space? What is the chronological antecedent of our concept of time? The chronological antecedent of the concept of cause? What is the logical antecedent of the concept of body? The logical antecedent of the concept of event? The logical antecedent of the concept of effect? The logical antecedent of the concept of a property? Give the table of the concepts which constitute phenomena, and the concepts which constitute noumena. Give the characteristics of phenomena, and also those of noumena. What is the rapidity of our mental operations when running in well-worn grooves? The rapidity of our spontaneities? The rapidity of our intuitive operation? The presence of the necessary concepts implies a faculty which produces them. The feebleness of intuition in the infant and the savage. The faculty of intuition derived from inheritance, and invigorated by discipline. Education of intuition must attend that of every faculty. What we know of necessary truths as forms of existence. Give examples.

Chapter XV.

EDUCATION—WHAT IS IT AND HOW ATTAINED?

194. What We Have Learned.—We have learned (Chap. II.) that the indivisible force called mind, manifests itself in three great classes of phenomena, namely, knowing, feeling and willing.[1] We have also learned that, in the experience of every mind, these three phenomena follow an invariable order of occurrence. It is clear that aside from the bodily sensations, knowledge alone begets feeling and feeling alone in the form of desire, prompts the will to act. We have found, moreover, that knowledge consists in the products of intellectual faculties whose early activity follows likewise a uniform succession from the initial act of gathering concrete ideas to the final act of reasoning upon their relations to each other.

It has been shown further that the feelings which the varieties of knowledge awaken, constitute also a series[2] wherein the bodily appetites and feelings termed selfish have the earliest vigor and are followed successively as the mind develops, by the social affections and the esthetic, the moral and the religious emotions. Finally, it has been shown in detail that every feeling of whatever class naturally desires its own gratification,[3] and that the will which crowns the series of mental manifestations, has the power to select from among the desires thus present to the mind, that one which it will strive to gratify. Manifestly the de-

---

[1] Chap. II. 23.   [2] Chap. II. 48.   [3] Chap. II. 41.

sire so selected, constitutes the motive which influences the will to put forth effort. It is consequently clear that, except the spontaneities, the will determines, directs and controls all human actions. The will, therefore, as prompted by its motive, is the grand agent[1] by which the higher emotions are nurtured, the lower feelings kept in check, the thoughts directed to worthy objects, the mind stored with valuable knowledge, and so disciplined as to attain the purposes of a true life.[2]

195. Education: What Is It?—Education of the mind consists in training every faculty of the intellect to complete efficiency, in subjecting the appetites to habitual control, in developing the higher emotions, especially conscience, to a uniform ascendency, and in thus furnishing for the will such elevated motives as will constantly prompt it to wise action.

196. (1) Physical Education consists in so invigorating, by systematic exercise and right living, all the organs of the body that each shall perform its complete function in harmony with the entire organism.

197. (2) Moral Education[3] consists in restricting the appetites and the selfish desires to the moderate indulgence which is essential to physical soundness and self-protection; in training to a permanent supremacy the social sympathies and the emotions of beauty, duty, justice, and right, and in so regulating the impulses of the will that they are habitually in harmony with a "conscience void of offense towards God and man."

198. (3) Intellectual Education.—The discipline of the intellect and the development of the higher

---

[1] Chap. II. 45.   [2] Chap. II. 47.   [3] Chap. II. 47.

sensibilities and the will, are inseparable in practice, but may be distinguished from each other in theory by definitions. Thus intellectual education is the harmonious training of all the intellectual faculties by successful efforts[1] and their reiteration until their united action in the attainment of knowledge, is comprehensive, rapid, accurate and facile.

199. (4) The Hand[2] which is the sense of resistance must be trained to delicacy of touch and facility of execution by practice in the manual arts especially those that afford the means of common expression; the eye as the sense of sight, must be trained by systematic exercise on visible objects, until it attains the power of rapid and accurate discrimination in the countless distinctions of color and form. The ear as the sense of hearing, should be cultured by repeated acts of listening until it reaches infallible exactness and ease in discerning their minutest variations. The memory[3] should be wrought to such readiness and tenacity by supplying it with clear and complete concepts gained through perfect action of the senses and other faculties, that it will retain and restore vividly to consciousness, every item of useful knowledge committed to its keeping.

200. The Faculty of Conception[4] must be disciplined by repeated efforts in picturing and realizing the concrete and other concepts recalled from memory, until it grasps and holds with facility and completeness, every concept that the mind contemplates. The power of analysis,[5] by means of protracted efforts in analyzing the concrete and complex ideas presented by

---

[1] Chap. IV. 75.
[2] Chap. III. 70.
[3] Chap. V. 82.
[4] Chap. VI. 111.
[5] Chap. VII. 121.

conception, should gain finally the habit of discriminating with exhaustive precision, the elements of a complete whole however intricate. The faculty of abstraction,[1] through reiterated acts of comparing the identical qualities revealed by analysis in concrete wholes, must reach at last a range of infallible action that renders the abstract concepts of properties held by memory equal in number and clearness with the individual concepts from which they were derived.

201. Imagination,[2] the Concept Builder, should be strengthened and refined; (1) by supplying it with the complete products of conception, analysis and abstraction as the material with which it constructs new wholes; (2) by presenting as the models after which it fashions its products, the more striking concepts gained through the senses; (3) by the continual construction of image concepts[3] that harmonize with genuine excellence and beauty in the world of reality. It is only by fulfilling these three conditions, that imagination can be trained to the habit of combining the materials so furnished into new images that shall accord with the dictates of common-sense and good taste.

202. The Classifying Faculty[4] derives the discipline which renders its action both comprehensive and accurate ; (1) from the clearness of the individual concepts to be classified ; (2) from a complete knowledge of the resembling parts and properties on which they are classified ; (3) from the distinctness of the corresponding abstract concepts used as standards in the classifying process; (4) from repeated and long continued efforts in forming the complete and accurate

---

[1] Chap. VIII. 125.
[3] Chap. IX. 136.
[2] Chap. IX. 129, 130.
[4] Chap. X. 144, 145.

classes out of the material [1] thus supplied. As the result of strenuous exercise in the process of arranging such a single concepts in classes, the faculty employed attains, after years of practice, the power of supplying with facility and precision the perfect class concepts that are the valid objects of class judgment.

203. Judgment and Reasoning.—Thus furnished with the abundant and definite class concepts [2] which are the finished products of the classifying faculty, judgment,[3] by continually scrutinizing and affirming the relations of classes to each other, slowly acquires readiness in the range and correctness of its decisions; and the truths which it clothes in propositions,[4] constitute the only materials out of which reasoning evolves concepts that are beyond the limits of personal experience. Finally reasoning, which depends for its object on the valid products of all the preceding faculties, must derive the culture that renders its conclusions, comprehensive and trustworthy, from protracted practice in the exact processes, both of induction and deduction [5] in which the series of intellectual operations finally culminate.

204. Every Faculty or Group of Faculties must be trained to supply complete Materials for the Action of the Faculty that follows it in the Series.—The education of the series of intellectual faculties as described above, evidently demands for its completeness, that each faculty should elaborate such perfect concepts [6] as will incite its immediate successor to its most effective action. The acts of reasoning, for instance, will be wide in range and accurate in conclusion according as the class judgments which it com-

---

[1] Chap. X. 144. [2] Chap. XI. 154. [3] Chap. XI. 156.
[4] Chap. XI. 157. [5] Chap. XII. 160. [6] Chap. XIII. 173.

poses in the syllogism,[1] are comprehensive in extent and correct in detail. In other words, the reasoning power can not be disciplined to great effectiveness by severity of practice, except on the condition of the prior effective operation of class judgment that prepares its materials.

This manifest dependence of each faculty on the completeness of its objects[2] for the effectiveness of its operations, should have great weight in arranging our systems of education, both in matter and method. The fact is evident that careless perceptions result in a bad memory; that a bad memory restores only obscure concepts, that obscure concepts on being analyzed, yield only confused elements, and that confused elements under acts of desultory comparison, result in abstract ideas that are vague and dim. It is equally evident also that obscure and confused concepts, concrete, simple and abstract, constitute imperfect materials for imagination whose products, the image concepts, will consequently lack symmetry and vividness.

It is beyond all question, moreover, that obscure concrete concepts can not be clearly discriminated as the units of classification; that confused elements or characteristics make no valid basis for the classifying process; that vague abstract ideas are inadequate standards by which to discern the distinctions on which classes are formed and, therefore, that the class concepts which result will be, in like degree, imperfect and narrow. But it is not less manifest, finally, that imperfect and limited class concepts produce judgments that are feeble and fallible, and consequently the efforts of reasoning which follow, will be mere conjectures that elicit truth only by accident.

---

[1] Chap. XIII. 162.   [2] Chap. XIII. 172.

205. The Environment Supplies Primarily the Objects by which the Faculties are Incited to Action.—Every individual mind is in the centre of an environment which comprises the objects[1] on which its faculties instinctively act. Our environment embraces the things and events that permanently surround us and come within the range of our personal observation and experience. It includes all the products of nature and art within our reach; all the branches of study we are led to pursue; all the influences parental, social, moral, religious, that continually mold our characters. In short, the environment is the sum total of the forces that constantly play upon us and incite our faculties, first, to spontaneous action and, finally, to the strenuous efforts that secure their discipline.

206. Selection of Objects that Elicit Effort.—But of the vast number of objects in our environment that stimulate the senses to action, especially in childhood, by far the larger portion excite only feeble and flitting spontaneities, while a comparatively limited number arouse curiosity and awaken those more intense and lasting spontaneities that induce the repeated efforts resulting at last in culture. Such objects (studies, science, arts) arranged in an order wherein the simple gradually approach the complex; the concrete reaches the abstract by successive steps; the particular becomes the general through a regular progression, constitute the means of intellectual education as guided by the teacher especially in the primary training of the child.

207. Studies must be Selected and Arranged in a Series that shall accord with the Series of un-

---

[1] Chap. I. 9.

folding Faculties.—It is clearly the province of the experienced teacher to select and adjust courses of study for his pupils that shall harmonize with the invariable order in which the faculties are called into action. In the attainment of this purpose, he must fulfil two conditions which are equally important.

(1) For each faculty or group of kindred faculties, he must construct out of the materials which its appropriate objects afford, a series of studies which shall begin with the simplest examples and rise gradually to those that are more involved.

(2) He must arrange the several series thus adjusted for the training of single faculties, in an order that corresponds closely with the uniform succession in the growth of the faculties themselves.

Respecting the first condition, let us note the manifest principle in education that the series of efforts by which a faculty is-trained, should increase in severity just in proportion as the powers that put them forth, increase in strength and facility of action. Thus the hand (sense of touch) is primarily trained by the handling of the simplest regular solids and combining them into groups of growing complexity. Then follow a course of lessons with the pencil in which the series of figures to be drawn commence with the simplest and gradually progress as the hand gains expertness, to the more complicated examples. Next follows the forming with the pencil of the simplest and most familiar words with countless reiterations, until it is judiciously succeeded by the copying of sentences which culminates at last, slowly and gradually, in the power to express thought in written language. This glimpse of the advancing complexity in the earlier processes by which the hand is disciplined, is a valid example of the series of progressive exercises by which every intellectual faculty rises naturally from the feebleness of its first efforts to the fullness of its strength. One of the vital problems which the science of education has yet to solve, is how to adapt *with perfect precision in all its details* each one of these exercises to the needs of the particular faculty which it is intended to unfold.

As to the second condition, namely, that the several series adjusted for the training of individual faculties, must be arranged in an order that corresponds closely to the uniform succession of the faculties,[1] we will only add here that its urgency in all the grades of instruction, is so great as to justify us in devoting a future chapter to its full elucidation. Nothing can be of greater moment in a system of education than the arrangement of its courses of study in a perfect logical order which accords with the successive steps by which the growing mind advances.

208. Every Intellectual Faculty can be disciplined only by strenuous and reiterated Efforts of Attention[2] directed to its Object.—It is a fact well-known to educators, that feeble and desultory efforts do not strengthen the faculty that puts them forth. On the contrary, the mind which occupies itself exclusively with acts of such a character, finally reaches a condition of permanent imbecility in which vigorous thinking is impossible. Intellectual efforts that are frivolous and futile, naturally lead to permanent mental degeneracy.

On the other hand, the vigorous exertion of a faculty concentrated upon its own objects[3] in science or art, and sustained until it attains in full distinctness the knowledge it seeks, invariably strengthens the faculty

---

[1] Chap. XIII. 173.   [2] Chap. I. 10.   [3] Chap. I. 9.

by which it is made. And such exertions, if unvaried and habitual, finally result in disciplined powers. For education as an attainment is the power which springs from strenuous efforts habitually and effectively expended upon appropriate objects.

209. Reiteration of Strenuous Efforts begets finally the Habit of Attention which results in culture.—The habit of earnest and exclusive attention to the objects of study and research, is, as we have frequently shown, the only means of gathering complete concepts and giving validity and precision to the processes of thinking that follow. The habit of perfect attention can be formed only by the *reiteration* of strenuous efforts of attention on identical lines of study. The knowledge which is most available in practical life and the intellectual operations which are most rapid and effectual, are the products of attention as strengthened by reiterated efforts. The habit of concentrated attention thus gained, is not only the *product* of reiteration but the *means* of continual attainment both in discipline and knowledge. Every expert whether in science or art, owes his perfect mastery of its principles to relterated acts of attention to all its details. Proficiency in any branch of study pursued by the pupil, is attained only on the rigid fulfilment of these same conditions.

It is then the manifest duty of the teacher in applying this principle of mind growth, to take care first that the pupil gains a perfect knowledge of each step in the subject he studies and, secondly, that he reiterates the mental process by which each step was mastered, until his mind attains spontaneous readiness therein. In the language of the school room it is first the perfect lesson and then the perfect reviews that discipline the intellect and supply familiar and practical knowledge.

210. Character of Concepts gathered by disciplinary Efforts.—The fact still to be recognized that, in the process of education, half knowledge is positively harmful both to the intellect and to progress in every particular enterprise. It is only clear and complete ideas, known in all their wider relations and narrower details, which attest the culture of the faculties that gather them, and assure success in their application to practical pursuits. The knowledge that is most useful in the world, is, other things being equal, the most effective as the means of education.

QUESTIONS ON CHAPTER XV.

Give the three great classes of mental phenomena and the order of their manifestation. In what does knowledge consist? The feelings developed in a uniform series. Show the relation between will and desire. What are the offices of will as prompted by the highest motives? In what does the education of the mind consist? What is physical education? What constitutes moral education? What is intellectual education? Example of educating the hand as the sense of resistance; the eye as the sense of sight; the ear as the sense of hearing. Example of training the conceptive faculty; the power of analysis, and of abstraction. Example of invigorating the imagination by fulfilling three conditions. From what three conditions does the classifying faculty derive its discipline? How does judgment acquire readiness, both in range and correctness? For what purpose should every faculty be trained to elaborate perfect concepts? What effect have careless perceptions on the memory? Give the character of the concepts which a bad memory restores. Analysis of such concepts produces what kind of elements? Character of the abstract ideas resulting therefrom. Imperfect materials for imagination produce what kind of image concepts? Imperfect and narrow class concepts result from materials furnished by the preceding faculties. What is the character of the judgments, and the reasoning processes which act upon crude and narrow class concepts? Relation of the environment to the action of the faculties. The selection of objects that elicit disciplinary efforts. Selection for the training of each faculty. Adjustment of the series for the successive faculties. Give examples. What is one of the vital problems in the science of education? How is every intellectual faculty disciplined? Effect of feeble and desultory efforts. Effect in education of vigorous concentrated efforts. Effect of reiteration upon vigorous efforts. Duty of the teacher in applying this principle. Effect of half-knowledge in education. The value of clear and complete ideas.

Chapter XVI.

THE SPECIAL MEANS OF TRAINING EACH FACULTY IN THE ORDER OF ITS GROWTH.

211. Principles in Education.—In presenting systematically the special means by which each faculty or group of kindred faculties may be most effectually trained, I shall adhere strictly to the following principles in the science of education.

(1) Every system of education must be based upon the laws of growth[1] both of the single faculty and of the succession of faculties that constitute the intellect.

(2) In training each faculty, the objects that stimulate it to action, must be arranged in a series which begins with the simplest and gradually advances to the most complex.

(3) In educating the intellect, the order of studies must be arranged in courses that accord with an invariable succession in the growth of its faculties.

(4) Such an arrangement of studies will present the whole before its parts;[2] the concrete before the abstract;[3] the individual before its class;[4] facts before fancies;[5] classified knowledge before definitions;[6] general propositions before reasoning.[7] Especially will this principle be admitted when we consider the abstract is derived from the concrete; that the class is composed of individuals; that fancies are built out of facts; that definitions depend on related classes, and

---

[1] Chap. XIII. 172, 173.   [2] Chap. VII. 124.   [3] Chap. VIII. 127.
[4] Chap. X. 145.   [5] Chap. IX. 130.   [6] Chap. X. 150.
[7] Chap. XII. 161.

that reasoning proceeds by means of general propositions.

(5) Every faculty is effectively trained (1) by an effort that exhausts the object and masters the process presented, and (2) by a reiteration of such efforts until it attains spontaneous facility.

(6) The efforts which educate a faculty should correspond in severity with the stage of its advancement, employing effectually but not exhausting its natural vigor.

(7) In early education only those objects of attention and study should be presented to the child, that excite his interest, awaken spontaneous action and stimulate effort.

(8) From exercises and studies that are equally effective as mental gymnastics, those should be given the preference that store the mind with knowledge that is practical and useful.

212. The Exclusive Training of a Single Faculty Impossible.—Thus equipped with the fundamental maxims of education, let us attend minutely to the special series of exercises and studies by which each faculty is educated. But let us first appreciate clearly the fact that, though science separates the faculties in theory, the exclusive training of a single faculty is impossible. For the strenuous acts of attention which discipline a single faculty inevitably gather the materials that incite the kindred powers to systematic action. Thus the act of drawing the form of a leaf, requires not only the careful movement of the hand, but the close attention of the eye by which the movement is guided, and the hand and the eye together furnish the concept of figure that supplies an object for memory, conception and analysis. The products of every faculty, as we have seen, especially when making disciplinary efforts, are the valid objects on which its immediate successor in the series expends its action. Hence the systematic attention of the senses each to its peculiar object, gathers concepts for all the processes of thinking that follow in the series.

213. The Sensations—How Trained.—The bodily sensations[1] and the appetites that attend them, springing, as they do, into vigor as the earliest spontaneities, need no formal training for the purpose of special development. Like the selfish feelings in general, they require only the uniform guidance and restraint which, confining them to proper objects, secure their legitimate purpose, namely, the safety and health of the body. Since the sole province of the sensations and appetites is to minister to the wants of the body, the training that adapts them precisely to this purpose, belongs to the department of physical education. It is mainly from his parents that the child learns the lessons of moderation and self-control in his sensual pleasures, that, taking the form of habit, promote the soundness and vigor of the body. Such are the conditions of early life that the teacher has comparatively little to do with the training of the animal senses, smell, taste, and sensitive touch. At most he can only act as the parents' assistant in seeing that his pupils refrain from any undue indulgence of their appetites while under his charge. An occasional precept on the necessity of self-control will be helpful in this direction.

214. Manual Training, the First Step in Education.—We have learned that the intellectual senses, touch, sight, and hearing, are the gatherers[2] of the concepts on which the other faculties expend their efforts. We have learned also that, in the order of its

---

[1] Chap. III. 58.   [2] Chap. III. 62, 63.

action, perceptive touch [1] precedes the sense of hearing and sight, and that it teaches the latter to distinguish form through the variations of light and shade that lie upon its surface. Perceptive touch, therefore, coming first into action, should receive the first formal training given by the teacher.

Perceptive touch, as we know, perceives solids simply through their resistance to the free motions of the hand. Touch distinguishes in bodies their impenetrability, shape, size, hardness, etc., simply as so many modes of resistance. The hand must, therefore, be trained by systematic contact with a great number and variety of resisting surfaces. The instinctive restless manual movements of the infant must be guided and gradually made intentional by furnishing it with solids suitable for handling.

From ordinary facility in discriminating differences in size and figure, the hand must be trained by methodical practice in a series of exercises wherein the perception of resistance increases in delicacy, step by step, until it attains the highest manual skill and expertness. But perceptive touch not only gives us our first ideas of solids in the outside world, but it becomes, as education advances, one of our most effective means of expression. It is in fact the main instrument by which the human mind embodies its best ideas. Without the help of the hand, the arts of writing, drawing, painting, sculpture, architecture, and the industries on which civilization is based, could never have had their birth. Thus it expresses [2] in colors and forms the thoughts that contribute most effectively to the progress of the race. Though less rapid and facile, its language is far more distinct, impressive and beautiful

---

[1] Chap. III. 70, 71.   [2] Chap. IX. 136.

than that of the tongue. And it is for this reason that the processes by which the hand is thoroughly disciplined, supply the mind with those concepts which induce precision of thought and effectiveness of expression. The Apothegm of Bacon hits this peculiar distinction between perceptive touch and the other intellectual senses as the agents of mental discipline. " Reading maketh a full man; conference a ready man, and writing an *exact* man."

215. The Hand then should be Trained.—(1) As a perceptive sense to discriminate with the utmost delicacy and facility of touch, the tangible qualities of the innumerable forms with which it comes in contact.

(2) As an organ of expression to delineate with ease and exactness any concept which the mind desires to represent.

Evidently the hand as a perceptive sense can be educated by a series of systematic manipulations on a succession of objects that begin with the simplest regular solids and advance gradually to the more complex.[1] The hand is also prominently active in the gymnastic exercises that are our means of physical education. Manifestly the hand both as a perceptive sense and as an instrument of expression, can be trained to delicacy of touch and skill in execution, by the same exercises. Thus, moulding, drawing, writing, spelling with the pen, and the use of tools in construction, if systematically, give finally to the hand automatic swiftness, not only in gaining ideas, but in committing them to language.

216. The Sense of Sight Trained by Exercises upon Visible Objects judiciously selected and arranged.—It is clear that the eye[2] participates in

---

[1] Prin. 2.   [2] Prin. 3.

every process by which the hand is formally trained and that, as a perceptive sense, it soon outstrips the hand in rapidity and range of discernment. But soon after the earliest lessons in manual training are commenced, the eye should be taught the distinctions of colors arranged in a series adapted to the purpose. First the color and next the shapes it indicates is nature's order. For color and shape—the first immediately and the last mediately—are the only properties which are discriminated by the sense of sight.

Exercises then on the innumerable variations of color, light and shade, and the shapes they indicate, furnishes the exclusive means of educating the eye. The sense of sight, from the dim and flitting perceptions of infancy to the full maturity of its power, should be trained by strenuous and accurate practice on objects that present, in the happiest combination and arrangement, the endless diversities of color, figure, size, distance and kindred qualities.[1] It is indispensable to the effective education of the eye that its objects, especially for early youth, should be arranged in a series which, commencing with the simplest examples of color and proceeding with color and form united, rises gradually to their more intricate combinations.[2] The adjustment of visible things to be used for the formal discipline of sight, ought to be adapted to the different stages of growth from infancy to maturity. And since form, which also is one of the qualities addressed to sight, constitutes the exclusive quality that appeals to touch, it is manifest that these two senses must be educated by simultaneous exercise on the same objects. In view of these facts it is clear that a philosophical course for the development of

---

[1] Chap. III. 71.   [1] Prin. 2.

sight and touch, can be arranged only by those who are acquainted with infant psychology on the one hand, and with the sciences of color and form on the other.

217. The Sense of Hearing Trained by Listening to Significant and Musical Sounds.—The ear[1] may attain finally great rapidity and acuteness of perception by strenuous attention given to exercises in both articulate and musical sounds. The earliest lessons for training this sense are the mother's voice and the child's own tongue. If the tongue be silent through any defect, the ear, as the organ of perception, is educated with great difficulty; while in cases of absolute deafness, the tongue is hopelessly mute.[2] The early efforts in learning to talk; careful phonetic exercises in school; the singing of simple melodies; conversations; the reciting of suitable verses; reading and elocution, all advancing step by step, to higher complexity and final completeness, constitute the principle means of educating the sense of hearing.

As in the training of every other faculty and especially the senses, great pains should be taken in these exercises to master every step before proceeding to the next one. The tongue gains the habit of distinct and felicitous utterance and the ear delicacy and quickness of discernment only by years of assiduous practice. Interest and reiteration are the indispensable means of making such practice effective. These two stimulants to thoroughness should go hand in hand.

218. The Training of each Intellectual Sense Stimulates the other two Senses to Disciplinary Action.—All the exercises that give facility and expertness to the movements of the hand, also stimu-

---

[1] Chap. III. 72.   [2] Prin. 2.

late the eye to corresponding action, and the clear notions gathered by the harmonious action of sight and touch, elicit naturally such accurate expression as help to train the sense of hearing. The action of the intellectual senses upon an object either separately or in concert, produces the effect on the mind known as a percept. Thus if for the first time, I examine, both by touch and sight, any object, say a new variety of the orange, and hear its name, the percept resulting will be composed of the elements gathered by the eye and the hand, closely associated with the name as addressed to the ear. Remove the orange and the percept becomes a concept,[1] which is full and distinct in proportion to the fullness and distinctness of the percept it represents. This concept is acquired by memory which retains it in unconsciousness until it is recalled to supply the material for subsequent thinking.

219. The Memory Trained by What Means?— Memory,[2] as we know, is the depository of related concepts. Whenever a vivid concept disappears from consciousness, it is acquired by the memory and held until wanted for further mental operations. Thus memory gathers in the order and under the relations of their acquisition by the mind, all the clear and complete products of every faculty. It is only the dim and imperfect concepts that fade out and leave no trace behind. Especially in the period of childhood does the memory acquire and preserve the distinct concepts gained by the strenuous action[3] of the intellectual senses.

220. Memory Trained by Vivid Concepts.—Since the three successive steps in memory of acquiring, retaining, and recalling concepts, are both unconscious and spontaneous throughout, this faculty cannot be

---

[1] Chap. XIII. 173. [2] Chap. V. 82. [3] Chap. V. 84.

trained like the other powers, by intensifying its processes through efforts of the will. The effectiveness of the operations in memory depend solely upon the character of the concepts which other faculties[1] furnish it. The most effective means then, of training memory to habitual readiness and tenacity, lies in the prior training of the faculties that supply its material.

221. Exercises that Train the Senses, Train the Memory also.—Since the percept which is the object of memory depends wholly for its completeness on the strenuous action of the senses, it is manifest that any exercises that train the senses, will train the memory also. Early memory then is developed by the systematic studies that train the senses of touch, sight and hearing, and later memory is disciplined by the earnest application of the mind to studies that train the higher faculties of the series to a power of action that is habitually effective. For, as we have already said in effect, it is the action of educated faculties that furnish the complete ideas which render the memory ready and reliable.

222. Interest and Reiteration.—The three operations of memory are, moreover, invigorated by the same conditions that incite to strenuous effort the faculties which supply its material.

Prominent among these is interest and reiteration. The feeling of interest[2] in the object of any faculty arising from its usefulness, beauty, novelty, value, and a reiteration of the processes by which an idea is gained therefrom, stimulates in like degree the faculty employed and the memory which stores its product. But in proportion as the interest is strong and absorbing, will the necessity for formal reiterations be de-

---

[1] Chap. V. 95.   [2] Chap. V. 86, 87, 88.

creased and vice versa. When the love of knowledge, for example, is eager and lasting, the concepts gained under its impulses, recur spontaneously and the need of many formal reviews is less urgent. On the other hand, objects that are viewed with a natural indifference, such as the mere inflections of a foreign tongue, the proper names of persons that are comparative strangers, or the particular dates of distant events, are fixed in memory only by frequent repetition.[1] The value of special Mnemonic exercises lies in the help they give in acquiring permanently ideas that are either unattractive in themselves or only distantly related to ideas already acquired.

223. The perfect Classification of Concepts help their Retention in Memory.—One of the values of the sciences when studied as the means of discipline, consists in the fact that the perfectly classified concepts[2] they present, are when thoroughly acquired; permanently fixed in the memory. For not only do strenuous attention, interest and reiteration tend to arrest the fading process, but perfect classification also helps to give fixedness to ideas that are really worth the gathering.

224. Conception: How Educated.—It is the province of conception[3] to realize the concepts that are recalled from memory. This faculty is simply the power the mind has of concentrating its attention upon a conscious idea and thus making it distinct and vivid. Its office is to represent clearly the products of all the other faculties. Conception grasps and gives conscious distinctness to an idea, while another faculty modifies and changes this idea into its own peculiar product. I cannot analyze for example my concept of

---

[1] Chap. V. 93, 94.   [2] Chap. V. 103, 104.   [3] Chap. VI. 111.

Westminster Abbey without first distinctly realizing it as a whole through conception.

There are then two obvious means by which the faculty of conception may be disciplined to habitually rapid and effective action.

(1) By supplying it through the effective efforts of the other faculties with concepts that contain all the elements of the thing they represent.

(2) By strenuous and repeated efforts to grasp and intensify each concept which is recalled from memory.

The first condition is fulfilled by educating the faculties that furnish the objects on which conception centers, the second by a concentrated attention upon every concept they present. Of course the order of the concepts (studies) presented for discipline of the conceptive faculty, will accord with principle (2.) It must begin in infancy with spontaneous action on the concepts of things that are simplest and most attractive. It must rise gradually in complexity as the mind advances towards maturity. In early education the order of studies that train the senses most effectively, will supply also the concepts that discipline the juvenile conception.

225. The Concepts and Processes that train the Faculty of Analysis.—In the order of nature, thought advances from the perception of a whole to the scrutiny of its parts. It is the province of analysis[1] to discriminate, one by one, the parts and properties of a concept which conception has realized as a whole, and thus to center the attention successively upon every element the whole contains.

The faculty of analysis is trained to habitual accuracy by persistent efforts that note with precision all

---

[1] Chap. VII. 121, 122.

the elements which the idea under scrutiny comprises. The effective discipline of the analyzing power then depends on two conditions namely (1) a series of concepts that are clear and full (2) a corresponding series of analyzing efforts that are strenuous and exhaustive. The concepts gathered by analysis, namely, the notions of the parts and properties discerned, should be as distinct as the whole they constitute. The early exercises that train the senses, evidently supply, in logical succession, the concepts for systematic training of early analysis. The object lesson makes the first formal demand for incipient efforts. Then follows the processes of drawing, reading, writing, spelling, arithmetic and botany, in which careful discrimination of the parts of wholes whose complexity continually increases, is constantly demanded from the pupil.

226. The Training of Abstraction: How Conducted.—The faculty of abstraction,[1] as we have seen, compares each property revealed by analysis of a concrete concept, with its duplicate as found in other concepts, until the mind reaches an idea of such property apart from any individual instance. The multitude of adjectives in language which designate the properties of concrete things, are the objects which the faculty of abstraction works up into subtle notions called abstract nouns. Abstraction likewise withdraws each element of thought or feeling from the concrete group wherein it manifests itself, makes it a separate concept and names it as a distinct entity. Thus (the adjective in the phrase) "an upright judge" expresses a concrete quality, while the word uprightness, denotes the abstract concept derived from it.

Manifestly the faculty of abstraction is educated (1)

---

[1] Chap. VIII. 125, 126.

By the uniform distinctness of the individual properties which are the objects on which it acts.

(2) By formal and reiterated acts of attention to the steps whereby the notion of each individual property is transmuted into an abstract concept.

(3) By realizing with increasing clearness the ideas presented by the abstract sciences.

The first condition is answered by the trained and perfect action of the faculties that precede abstraction in the series, particularly that of analysis. Especially in early education, the senses must gather concepts that comprise all the characteristics of the things they duplicate. The memory must retain and restore these concepts in undiminished fullness, conception realize them with exclusive vividness, and analysis disclose with equal clearness each characteristic they display. The second condition is realized by the careful and systematic study of concrete units that call the attention of the pupil to the qualities of number and form. The third condition is fulfilled by realizing with a distinctness that increases with incessant repetition, the ideas which the abstract sciences present to the faculty under consideration. Such are the sciences of arithmetic, grammar, geometry, etc.

227. Ways and Means of Educating Imagination.—Imagination,[1] as we have learned, is the faculty which combines into new wholes the concepts supplied by the three preceding faculties, namely, conception, analysis and abstraction. Concrete concepts, concepts of their elements, and abstract concepts are then the materials[2] out of which imagination constructs new images that have no external counterparts. The growth of imagination may be divided into three periods.

---

[1] Chap. IX. 129. [2] Chap. IX. 130.

(1) The infant period which spontaneously moulds into grotesque images the concepts of sense that are imperfectly analyzed.

(2) The juvenile period wherein the imagination combines into more consistent wholes the individual elements (properties and parts) which analysis has more fully disclosed.

(3) The period of maturity when the three preceding powers have supplied, through their complete and harmonious efforts, a wide range of abstract concepts out of which imagination constructs those symmetrical images that a cultured taste can commend.

228. The Special Method of Training Imagination.—Since the acts by which conception, analysis and abstraction produce each its peculiar concepts, supply the materials out of which image concepts are constructed, it is manifest that the antecedent training of these three faculties is essential to the right training of the imagination. For since the action of this faculty is mainly spontaneous, its culture will depend largely on the completeness of the concepts that constitute its materials. Moreover, the concrete concepts are in some sense the models after which the image-making faculty fashions its creations. Hence imagination is developed by fulfilling two conditions.

(1) By presenting the exact materials that stimulate imagination to facile action.

(2) By studying constantly the most genuine products of nature and art.

The first condition has been made sufficiently clear; the second condition will be fulfilled by the training of the senses to gather the most striking ideas[1] of things, especially by practice in the manual arts and studying the more simple products of imagination in

---
[1] Chap. IX. 135.

literature. Color studies, drawing, writing, composing, constructing with tools, studying language and many other pursuits, constantly collect the concepts that incite this important power to its appropriate action.

229. The Classifying Faculty: How Educated.—The classifying faculty, as previously defined,[1] is the power the mind possesses of gathering into systematic groups, those individual concepts that have resembling characteristics. The adequate preparation of the materials for the classifying process, requires three antecedent serial operations that rea accurate and complete.

(1) The disciplined action of the conceptive faculty producing concrete concepts that are distinct and full.

(2) The precise action of analysis that reveals exhaustively their characteristics (properties and parts).

(3) The valid efforts of abstraction which evolve the general ideas of which the mind discriminates properties in the concrete. (See Principle 4).

The severe training of these preceding faculties will, therefore, secure the finished materials that stimulate the classifying faculty to complete and effective action.

But classification is educated most effectively by systematic efforts in arranging concrete concepts[2] into well defined classes. The first formal training begins with systematizing the spontaneous acts which form rudimentary classes. The first lessons in counting are efforts in classifying, and the child can hardly form a simple sentence without using the classifying process. The elementary steps in the sciences stimulate classification to exactness of arrangement. On the other hand, the desultory classifying[3] of the ordinary objects

---

[1] Chap. X. 145. [2] Chap. X. 157. [3] Chap. X. 146.

that meet the eye, has no disciplinary effect. Extended practice in studying and classifying the simplest objects in botany, zoology, or mineralogy would certainly help to systematize the child's habits in the classifying process; while the subsequent arranging of groups within groups in the ascending series of classes which every science presents, would give his mind facility and precision in the more difficult examples. The careful study of the related classes that serve as the bases of definitions,[1] contribute to the same result.

230. The Concepts and Processes that Educate Class Judgment.—Remember that class judgment[2] is the faculty which, comparing two class concepts, affirms that one is, or is not, contained in the other class judgment[2] perceives the harmony or the discord between two concepts under scrutiny and when in harmony affirms that one *includes* the other as a subordinate class and, when in discord, that one *excludes* the other as a subordinate class. The material then on which class judgment operates is not singular but dual. Two conditions will, when uniformly fulfilled, secure the effective action and final discipline of class judgment.

The first is the perfection of the concepts under comparison. Evidently the act of comparison can be exhaustive and certain only on the condition that the mind has previously acquired a clear and comprehensive knowledge of the essential elements contained in the concepts which the judgment compares. But such knowledge can be obtained only through the perfect classification of which they are the products. The facility and precision of the innumerable acts by which judgment is cultured, consequently depend, apart

---

[1] Chap. X. 150.   [2] Chap. XI. 155.   [3] Chap. XI. 156.

from their own vigor, upon the efficiency of the preceding acts [1] of classification that prepares its objects.

The second condition on which judgment is educated, is the precision of its own affirmations. Nothing is so conducive to the attainment of a valuable judgment as accuracy of statement. To learn and affirm the actual relations of things of whatever class, supplies for judgment and reasoning, their perfect gymnastics. Every branch of study, whether of science or art, from beginning to end proceeds in judgments, and no instruction can be given except in propositions [2] that clothe the products of judgment in words. The means of its training are therefore ample, and the right arrangement of studies for early education and the love of truth throughout, will result finally in its genuine culture.

231. **The Means of Educating the Reasoning Faculty.**—We have already learned that the affirmations of class judgment are the initiatory steps in the process of reasoning.[3] For a single act of judgment consists in affirming that one class concept contains or includes another, while the simplest act of reasoning consists of three acts [4] of judgment compared in the syllogism.[5] The syllogism [6] then is the form, and the concepts of judgment the material, in the reasoning process; and since the materials for reasoning are the concepts of judgment, the accuracy of the conclusions it reaches, will, aside from the process be determined by the accuracy of judgments that precedes it. It is consequently clear that the genuineness of its material and the precision of its efforts, are the means by which the reasoning faculty attains finally the ability to con-

---

[1] Chap. XI. 154.
[2] Chap. XI. 157.
[3] Chap. XII. 159.
[4] Chap. XII. 161.
[5] Chap. XII. 162.
[6] See Syllogism, Chap. IX.

duct, with logical acuteness, the most complicated processes of thought in search of truth.

Since the operations of judgment and reasoning are the highest forms of thinking, it is the purpose of a systematic education to give these faculties their highest culture. The means are abundant. The mathematics whose examples are the processes of perfect logic, grammar, history and the natural sciences generally, are among the studies that incite to their most strenuous efforts, the powers in whose exercise human thought reaches its culmination.

QUESTIONS ON CHAPTER XVI.

Give in their order the Principles in Education (1) (2) (3) (4) (5) (6) (7) (8). Why is the exclusive training of a single faculty impossible? What is the proper treatment by the teacher of the bodily sensations of the pupil? What intellectual sense is first in action? By what means does perceptive touch perceive solids? By what means then should the hand be trained? Explain the office of the hand as an instrument of expression. What arts depend upon the hand for their development? What is the difference between the language of the hand and that of the tongue? What are the special means of training the hand as an instrument of expression? In what order should these means be arranged? When should the training of the eye be commenced? By what means is the earliest training of the eye given? How should the objects used in training the eye be arranged? Why are the eye and the hand trained by simultaneous exercises? Give the means of training the ear, and their arrangement. What are the two conditions on which the three intellectual senses are effectively disciplined? What determines the elements contained by the percept? Define memory. What is the character of the concepts acquired and retained by memory? Give the source of the concepts gained by the memory in childhood. Why can not the memory be trained like the other powers by intensifying its processes? Give the most effective means of training the memory. Why does the training of the senses train the memory also? How is the later memory disciplined? What is the effect of interest and reiteration on memory? Why does increase of interest diminish the necessity of reiteration? Value of mnemonic exercises. The effect of perfect classification upon retention in memory. What is the office of conception? By what two means is the faculty of conception disciplined? What arrangement of studies will best discipline the conceptive faculty? What is the province of analysis? On what two conditions does the discipline of the analyzing faculty depend? What are the early exercises that train the faculty of analysis? What is the process of abstraction, and what its products? Give the three means by which the faculty of abstraction is educated. Explain how the first means is used to accomplish its object. Also the second. Also the third. Define imagination and give the three periods of its growth. What faculties supply the materials with which imagination builds its images? On what does the culture of imagination depend and why? Give the two conditions on which the development of imagination depends. How will the second condition be fulfilled? Define the classifying faculty and show from what three preceding operations it gains its materials. On what three conditions does the completeness of these operations depend? Classifying educates most effectively by arranging complete concepts into classes. With what exercises does the early training of classification commence? Give the effect of desultory classification. Define class judgment. On what concepts does class judgment operate? What is the first condition on which class judgment an be disciplined? What is the second condition, and how fulfilled ! What are the concepts on which the reasoning faculty acts as its objects? What three acts of comparison constitute the reasoning process? On what will reasoning depend then, for the accuracy of its conclusions? What are the means of educating the reasoning faculty, and how arranged?

Chapter XVII.

EXPRESSION AS A MEANS OF INTELLECTUAL DISCIPLINE.

232. Expression is the act or acts by which the mind makes known to others, its own ideas, feelings and volitions. The various means whereby a thought, feeling or purpose manifests itself to the senses of those around us, are comprised in the comprehensive term *language*. Language is of two kinds, namely, natural and artificial language. Natural language embraces all those modes of expression which the mind is prompted by nature to use without the aid of formal instruction. Artificial language consists in the knowledge and employment of artificial signs in communicating ideas, emotions or desires.

233. The Means of Expression in Natural Language.—The natural language used in common by the entire family, civilized or savage, is composed of gestures of the hands, movements of the body, modifications of the face, and of a variety of inarticulate sounds uttered by the voice. Natural language is instinctive and spontaneous throughout. Like all the instinctive spontaneities, it is inherited and develops itself without the help of formal training. Every passion, emotion, desire has its peculiar mode of manifestation. Anger has its frowns and threatening gestures; fear its shudder and its scream; grief its tears and sobs; pain its groans and cries; joy its smiles, and the visible emotion its peals of laughter. The pure interjections and all inarticulate cries are natural means by with each feeling and especially every passion expresses its intensity.

234. Character of the Sounds in Natural Language.—The painful feelings, especially when vehement, express themselves in harsh and unmusical sounds while the pleasant feelings, especially the higher emotions, find utterance in tones and inflections of voice that are uniformly melodious. Upon the countless modulations of voice that express, in natural language, the multitudinous shades of the human sensibilities, are based the arts of elocution and music. These two arts which, have been developed to high excellence by the study and practice of experts throughout the world, constitute important branches in artificial language.

235. Judicious Training in Expression by Means of Natural Language.—Since the utterances of natural language are, like the feelings they express, wholly spontaneous, the training they receive should be such as to establish the habit of moderation and judicious self-control. Especially will a well considered system of early instruction teach the child to check habitually the expression of the lower feelings which are naturally violent, and to utter without restraint the kindlier emotions that are dawning within him. For, as we shall see more clearly farther on, every feeling, passion or emotion is strengthened by expression or restrained by silence.

236. The Means of Expression in Artificial Language.—Artificial language employs, as a means of expression, signs that are wholly conventional or natural signs that have been improved and rendered effective by art. The latter comprise music and elocution; the former, speech, drawing, painting, writing, and constructing.

(1) Music is the art of expressing the higher emotions, especially those of love and beauty, by means of melodious and harmonious sounds.

(2) Elocution is the art of impressing thought and feelings upon others, by uttering words with fitting distinctness, emphasis and suitable inflections of voice.

(3) Speech is the art of communicating thought and feeling by the vocal utterance of significant sounds or words.

(4) Drawing, as an art of expression, portrays an object or concept by means of lines drawn with the crayon or the pencil.

(5) Writing is the art of representing by visible characters made with the pen, the words of spoken language and thus expressing to the eye the thought they embody.

(6) Painting is the art which represents to the eye, by means of color, the concepts of forms especially those that display the element of beauty.

(7) Constructing includes the various arts and handicrafts wherein solid materials are so combined or modified as to express ideas previously conceived. Constructing includes, in the fine arts, sculpture and landscape gardening; in the useful arts, all the branches of industry that work up raw material into products that conduce to the convenience and comfort of mankind.

237. Artificial Language addressed to the Ear and the Eye.—Manifestly speech, music, and elocution, which includes audible reading, are addressed to the ear, while drawing, silent reading, writing, painting and constructing, are addressed to the eye. Consequently, as we have already seen, the former afford effective means for training, through acts of attention, the sense of hearing; the latter for training the sense of sight.

238. The Instruments of Expression in Artificial Language.—The organ of expression in music, elocution, and speech, is the tongue, while the hand as guided by the eye, is the instrument employed by the mind to embody its ideas in material forms. The tongue then is trained to facility and accuracy by acts of careful utterance in music, elocution and speech. The hand attains skill and precision by the actual making of the forms in drawing, writing, painting, and constructing, that represent distinctive ideas. Thus they exemplify the maxim in education that "we learn to do by careful doing; to think by careful thinking."

It is evident, as noticed above, that, in all the acts of expression, the products of the tongue are the objects of the ear, while the products of the hand are the objects of the eye. In this way the tongue and the hand under effective training, as instruments of expression, furnish the natural means for educating the senses of sight and hearing. In other words, all the methods of artistic utterance and all the products of manual skill are available in imparting swiftness and delicacy of discernment to these two feeders of the intellect.

239. The Value of Spoken and Written Language as Gymnastics for the Intellect.—The disciplinary effect of practice in the manual arts, has been explained elsewhere under appropriate heads. It remains to show the importance of language, spoken and written, as an auxiliary in educating the intellectual faculties. The value of study and the careful practice in language as a means of discipline, is based, if I mistake not, on the following facts.

(1) No concept of whatever kind is complete until

united with a name. The concept and its name are not two different entities. They are rather inseparable constituents of a compact whole. The name minus the concept, is a nullity; the concept minus its name, is vague and rudimentary. The truth is, the notion of a significant sound that clings to the concept as its sign, is one of its essential characteristics. Lacking this characteristic, the concept lacks also the clearness which is indispensable to valid thinking.

(2) The word that designates a concept, answers two purposes, namely, to give it vividness in the mind of the thinker and to serve as a means of expressing it to others.

240. Effect of Expression on the Concept further Explained.—Every thoughtful person has noticed that the effort of expression stimulates the processes of thinking. A speaker, for instance, begins his discourse with a slow and often painful enunciation, but, as he progresses, the words he employs impart gradually a glow to the thought they embody and, touching the numerous springs of association, soon attain facility of utterance through the clear and definite ideas they recall from memory. In perceiving a strange object or gaining a new idea, the first intellectual necessity is to learn its name. It is the name especially that enables the mind to give a new idea its fullness, to adjust it in its proper class, to retain it in memory under the close association of "sign and thing signified" and to express it clearly when recalled. This is the reason why the infant can retain and recall only the simple notions whose names he has learned and can pronounce. Without the power to express an idea the infant mind is unable to hold it with a definite grasp.

Further, all glowing and vivid ideas naturally seek

for expression; they rush impulsively to the tongue; they take to wing in language. And thus they not only from given utterance an increase of their own vigor, but find access to other minds and, if worthy of the distinction, become the common property of the race. These are some of the reasons why a knowledge of language and its laws, is indispensable to intellectual culture.

241. Every Faculty has a Language which is essential to the clearness of its products.—It will be found, on careful scrutiny, that each of the intellectual faculties has its own peculiar vocabulary, which is employed to express its action and the products evolved thereby. The verb *touch* or *feel* which affirms the action of the hand is followed by a variety of names (nouns and adjectives) that designate the properties of a solid, round, square, long, hard, sharp, are examples. As objects of the acts expressed by the verbs, see, view, observe, come a countless host of terms which denote visible things and their qualities. These terms are, for reasons we have already learned in Chapter III., largely identical with those employed by the sense of touch. In like manner the verbs that affirm the acts of hearing, have for their complements a multitude of words which specify sounds and their many properties and modifications. Manifestly the training of these three senses will be thorough in proportion as their action is uniformly strenuous and exhaustive, the resulting percepts complete, and the words that discriminate these percepts with the elements they contain, are fully mastered.

We have seen how the sign and the thing signified[1] are compacted together in every concrete concept

---

[1] Chap. V. 102

acquired by memory through the senses. We know also that this name not only imparts to its concepts the vivacity that prevents its fading, but serves as the means by which it is recalled from memory and presented with undiminished fulness to the conceptive faculty. Thus enlivened by its name, the concept as a whole is now subjected to the process of analysis, which, if minute, discloses completely its parts and distinguishes each *with a name*. The names of the parts discerned by analysis in a concrete concept, are *nouns*, as the stamens, pistils, petals, etc., of a flower; while the properties are designated by adjectives, as yellow flower, large flower, beautiful flower. It is clear that the analyzing process could never be carried beyond a vague spontaneity except as the elements it discovers are discriminated by names.

Further every property perceived by analysis is, by comparison with identical properties in other objects, withdrawn as a concept by the mind, from the identical concepts in which it exists, transformed into an abstract idea, and crowned with a name. This name is termed in grammar an abstract noun. But the abstract concept has no corresponding object in actual existence. Beauty, for instance, exists only as a property of concrete things, consequently, the notion of beauty in the abstract rests wholly upon the word that represents it. Lacking the name, there would be no such product as an abstract concept.

But imagination, in the absence of language, would have for its materials only rudimentary concrete concepts with the vaguest notions of their superficial contents. Consequently in the utter lack of abstract ideas, it would construct mere crude and grotesque images, whose means of expression would be as limited as that of the lower animals.

Again, without the language of words, classification beyond the narrowest limits, would be also impossible. For in the first place, its materials would, as we have shown, consist of a few vague, concrete concepts whose characteristics, confused and dim, would be attended by no abstract concepts as standards. Moreover, in every class arranged by the classifying process, the individuals it includes, are bound together by means of the class name or common noun. For instance, a countless host of resembling objects are grouped together under the term *tree*. Destroy the name and all means of designation, and you wipe out the class concept along with it.

Finally, in the deficiency of definite class concepts, judgment would have no ideas whose relations it could affirm. Especially in the lack of language, the proposition or sentence in which its affirmations are clothed, would be utterly wanting. As a consequence, the reasoning faculty, since it could have neither class concepts for its objects nor propositions with which to construct the syllogism that constitutes its process, would remain forever an undeveloped germ. Hence the human mind without the language of significant sounds, would have been incapable of advancement, and man would have held perpetually a grade of intelligence lower than that of the primitive savage.

242. The Comparative Importance of Language in Courses of Study.—Every series of intellectual exercises that result in the culture of the faculties employed, afford systematic practice in the use of language. But in view of its vital importance to mental progress and its value as a gymnastic, the study of language from its simplest form to its highest complexity, should occupy about one-third of the systematic courses arranged for general education. The or me minute features of language training are developed elsewhere especially in the last two chapters.

QUESTIONS IN CHAPTER XVII.

Define expression, and give the term that embodies its means. Define language both as natural and artificial. What are the means of expression in natural language? How does each feeling naturally manifest itself? How is expression by means of natural language judiciously trained? What are the means of expression in artificial language? Define each. To what senses is artificial language addressed? What are the organs of expression in artificial language? What effect has the act of expression in training the organ? What maxim in education do they exemplify? In expression what is the relation of the tongue to the ear, and the hand to the eye? Value of the tongue and the hand in educating the ear, and the eye. What is the value of spoken and written language as a gymnastic for the intellectual faculties generally? The effect of uniting a concept with its name. What two purposes does the word that designates a concept answer? Explain further the effect of expression on the concept expressed. What is the tendency of all vivid concepts? Show how every faculty has a language which is necessary to the clearness of its products. Give distinctively the language of the following faculties, and show the impossibility of their development without language. The language of touch, sight and hearing. Language as a means of retaining concepts in memory, and recalling them. Language of conception; of analysis; of abstraction, with the effect of its absence on each. Effect of the absence of language on imagination. On classification. Twofold effect of such absence on judgment. Also on the processes of reasoning. Condition of the human mind in the absence of language. What rank should language hold in courses of study? Comparative time that should be given to the study of language in our systems of education.

Chapter XXIII.

HIGHER SPONTANEITIES SPRINGING FROM TRAINED EFFORT CONSTITUTE OUR PRACTICAL KNOWLEDGE.

243. The Higher Spontaneities;[1] How Produced.
—But the strenuous effort that produces disciplined power, not only follows the more habitual intense spontaneities, but tends to become itself spontaneous when often repeated without variation. Any earnest effective intellectual act or series of acts, when frequently renewed, becomes, at each reiteration, less strenuous and more facile, until at last it reaches the perfection of a higher spontaneity. The most complicated and difficult process of thought begun and completed by concentrated prolonged exertion, requires, at each exact repetition, less application of will force, and so finally ends in swift and subtle spontaneities which need but a single act of the will to call them forth. In fact, the higher spontaneous operations which had their birth in strenuous effort, though they sometimes include several distinct steps, are completed often with such rapidity as to elude consciousness From this law of higher intellectual activity, comes the fact that the final and most finished acts of disciplined faculties, are subtle unconscious spontaneities which the will directs, not in detail, but in aggregation. The adept, the expert, the orator, each in his line, have entered the domain of the higher sponta-

---
[1] Chap. I, 7, V. 106, 107.

neities. Genius[1] is inherited proclivity for the higher spontaneities by whose flashes new truths are revealed.

244. The Will Directs the Higher Spontaneities in Groups.—When the spontaneous processes evolved from repeated, strenuous efforts, have become so rapid that the separate acts which compose them, outstrip the impulses of the will from which they originated, the mind thereafter directs by a single effort the entire group. In this manner, each act which was primarily a strenuous effort, is now automatic, and the will, by a single impulse, controls the group, while the separate successive acts it includes, are completed spontaneously.

Illustrations of this capacity of the intellect to quicken the processes of thought, which consist at first of strenuous efforts, until, by reiteration, they reach automatic rapidity, may be found in every one's experience.

245. Illustrations of the above Principle.—A familiar instance may be noted in the alphabet.

Whatever may be the method employed, the child learns the alphabet by successive acts of attention to the form of each letter. But frequent reviews gradually make the effort of attention to individual letters, unnecessary, until at last the will needs but to touch, so to speak, the initial "A" when the names and forms of the remaining letters flash through the mind in regular succession with automatic swiftness. Perhaps in most minds the pronouncing of the letter "A" for the purpose of recalling spontaneously the entire alphabet would be followed by the regular recurrence of the names only; but a slight modification of purpose will bring into consciousness automatically the forms along with them.

---

[1] Chap. IX. 141.

Another familiar illustration is found in the multiplication table, which is learned with laborious detail wherein the tyro dwells on each step in the different series with a distinct exertion of the will. But the effort of the will decreases and spontaneity increases inversely at each repetition, until the steps which constitute each set, attain an automatic celerity that brings them successively into consciousness, whenever a single act of attention is directed to the first step.

The spontaneous recurrence of entire poems when the first lines with which the stanzas following are associated in memory are repeated, are likewise instances of acquired spontaneity.

246. Example from Reading and Classification. —But the process by which the child learns to read, illustrates more happily perhaps than other examples, the transformation, step by step, of special efforts into subordinate automatic acts. The pupil begins by minute attention to each of the elementary forms that compose a syllable, then to each of the syllable forms that are combined to complete the word, and finally to the complex form of the word itself. Each one of these forms which are elements in the resulting compound, the eye, in a synthetic movement, gathers by a discriminating effort.

But, in elementary reading, the act of discriminating the forms of identical letters, is far oftener reiterated than the act of discriminating identical syllables; and the act of discriminating syllables is much more frequent than that of determining words. The consequence is that, as practice in reading progresses, the recognition of the letter-forms in a word becomes automatic proportionately sooner than the recognition of the syllabic-forms; while the perception of the latter reaches automatism earlier, by a similiar ratio, than the perception of the word they compose. In the end the pupil is pronounced a good reader, certainly so far as the eye is concerned, when he distinguishes, at a quick voluntary glance, the total word while, at the same indivisible instant, there takes place an unconscious spontaneous analysis of its elements, which in audible reading, is followed immediately by its pronunciation. Indeed, in the case of those who read with unusual facility, the spontaneities, through unlimited practice, encroach so far upon the field of conscious effort that the eye travels two or three words in advance of the voice, so that conscious perception, unconscious spontaneities and vocal utterance seem to occur simultaneously.

The complex act of adjusting an unknown individual in a familiar class,[1] whether of species, genus or order, is an additional instance of the swift unconscious movement of trained spontaneity. A single instance will suffice. If one catches sight for the first time, of a particular orange tree, he at once designates it by the name of the species to which it belongs. Yet before he can name it an orange tree, the mind carries through a series of unconscious automatic acts which, following a slight effort of will in sense-perception, closes with spontaneous classification. Or, to inspect the successive steps in detail, the series really opens with a spontaneous movement of sight caused by the image of the tree as depicted on the retina. This spontaneous movement is instantly succeeded by *an effort* of sight that deepens and completes the entire concrete concept, which in turn, by a swift reflex action, is analyzed and then, as guided by the abstract concepts corresponding

---

[1] Chap. XIII. 176.

to its particular characteristics, instantaneously adjusted in the class that represents its species.

In this series of intellectual acts three elements only are revealed to consciousness; namely, the initial spontaneity, the effort of vision and the final class concept. The concrete concept; the act and results of analysis; the abstract concepts involved; and the classifying act that completes the series, are all so subtle and swift as to escape consciousness.

The above is an instance of ordinary classification spontaneously reached by an ordinary observer. But the mind of the botanist who has frequently traversed, with scientific minuteness all the unvarying characteristics that distinguish the successive groups from variety to kingdom, to which the orange tree belongs, would complete the consecutive steps of a perfect scientific classification with a similar automatic celerity.

247. The Value of Reviews Illustrated.—The psychological law that habitual efforts of will directed in identical lines, culminate in automatic perfection, is so important in education as demonstrating the value of reviews, that a still further illustration may be profitably studied. In the oft repeated efforts or, in ordinary parlance, the assiduous practice by which the tyro in piano playing becomes an expert. The early lessons consist in painstaking endeavors, in which every touch of a key is prompted by a distinct act of the will and a series of careful volitions precisely corresponding to the series of musical notes which compose the piece selected for practice. But if the reviews of the same air are continued, the exact efforts succeed each other with increased facility and each note requires less and less exclusive attention; finally the will force is wholly withdrawn from the minor elements of musical sound that compose a passage, and the deft fingers directed by conscious effort only at its beginning, elicit, with automatic precision, the most difficult music.

248. Conclusions Reached.—A complete analysis through consciousness of the processes instanced above by which intellectual efforts, slow and laborious in the beginning, are changed by reiteration to facile spontaneities which the will thereafter readily controls *in sections*, warrants the following conclusions:—

1. That the processes of thought in a disciplined mind, whether in the line of literature, science or art, consist in large measure of trained spontaneities which have originated in severe exertions.

2. That in the movements of thought, all the high velocities are automatic acts that have individually outstripped the impulses of the will, which now directs the groups of which they are the elements.

3. That there are two great classes of spontaneities or automatic acts; namely, (1) the *instinctive* spontaneities that precede and awaken voluntary effort, and (2) the higher trained spontaneities that are the product of voluntary effort often reiterated.

4. That intellectual efforts severe, protracted and effective, are the only means of educating both classes of spontaneities. The first by the reactive force, the second by the results, *of direct repetition*.

5. That all the higher intellectual attainments must have risen to the quick facility of trained spontaneities before they are fully available in actual practice.

6. That in all processes of thought whereby new concepts are evolved, effort strenuous and prolonged is the sole factor, and spontaneity the result of subsequent reiteration.

7. That consequently there are two legitimate purposes to be sought in the attainment of intellectual discipline. First, to incite in the mind of the student, such severe and prolonged self-exertion as will enable him to reach and appropriate truths that are new to his experience; and secondly, to repeat these initiatory efforts until they become trained spontaneities.

The first eventuates in the power to pursue successfully original investigation. The second reduces both the processes and the results of such investigation to the familiarity of available knowledge.

It is scarcely necessary to add that the first purpose is accomplished by the efforts that conquer the *advance* lesson with the least possible aid from the teacher and by the strain of the original researches ôf whatever character. And the second purpose, as is equally obvious, is attained by frequent and thorough reviews of the processes inaugurated and the products gained by original efforts, so that the mind finally takes its acquisitions out of the range of mere text-book knowledge and makes them its own.

249. Maxims Derived from the above Facts.— Several important educational maxims are based upon the above facts respecting the origin and quality of our habits of thought:—

1. Since the higher spontaneities that constitute culture, derive their character from the strenuous efforts that give them birth, it is of vital consequence that these efforts should be accurate and complete in every particular.

2. It manifestly follows that in every school-room exercise thoroughness and accuracy have far greater value as factors in education, than any amount of knowledge however great wherein these factors are

wanting. It is of the utmost moment that the extent of every exercise in the school-room, should be gauged by the pupils capacity to compass it minutely and exhaustively.

Since spontaneities that are facile and correct are the offspring of *reiterated* efforts in which these qualities are prominent, it is important to the last degree, that all reviews should realize the highest standards of excellence and that they should be made at frequent intervals until permanent spontaneity is attained.

The preceding analyses of the processes of thought bring to light the two defects that often characterize our school-room instruction; namely, advance lessons half mastered through feeble and fitful efforts, and reviews that are superficial in quality and insufficient in number.

The consequence is that the sort of discipline gained is such as to confirm the habit of putting forth efforts that are languid and that the knowledge which results, lacking both original completeness and automatic facility, soon fades from the memory and disappears forever. The graduates of high schools, and even of colleges, that apply accurately the methods or remember minutely the facts gathered from the sciences they studied, are indeed exceptions.

QUESTIONS ON CHAPTER XVIII.

What is the effect of earnest efforts repeated without variation? How are difficult processes of thought rendered facile and swift? How does the will direct the higher spontaneities? Give an example of the exercises that employ the higher spontaneities successfully. Explain how a spontaneous knowledge of the alphabet is gained by the child. Give an example of spontaneous classification. How is scientific classification made spontaneous. Illustrate the importance of reviews by lessons given on the piano. Sum up the conclusions derived from the foregoing principles. Give the educational maxims based on the facts respecting the origin and culture of our higher spontaneities.

Chapter XIX.

INJURIOUS EFFECTS OF WRONG ARRANGEMENT OF STUDIES.

250. Adjustment of Objects to Faculties.—If the adjustment of elementary facts or rudiments of science employed to engage and train the youthful attention, is such as to call each power into exercise at the time when it is naturally and spontaneously active, the happiest results may be looked for. But when the materials for instruction are so unwisely chosen or so faultily arranged, as to appeal to faculties which are still too weak to yield a strenuous response, the result is mental confusion and apathy rather than effective attention and clean-cut knowledge.

Now the period in which a faculty is naturally and spontaneously active, depends, in general, upon its place in the series [1] and upon the materials prepared and presented by preceding faculties. For instance, the powers of abstraction cannot act with precision and fullness, unless the act of analysis has previously supplied the simple individual concepts out of which, by successive comparisons, it can evolve abstract ideas. To illustrate concretely—a young child will comprehend when he is told that, if he gives his dog a little of his own food, his dog will follow him, because the words *dog*, *food*, and *follow*, express concrete concepts that have been previously gathered by his own observation. But when the child is informed that "virtue is its own reward" the words impart no corresponding ideas, because of the

---

[1] Chap. VIII., 173.

fact that virtue is an abstract concept evolved from numerous similar acts by comparisons which he has never yet made. *Reward* is likewise a word whose significance may be dim for a like reason. Precepts of this character then, since the terms that embody them are meaningless, pall[1] on the attention of the early pupil and produce apathy if not aversion to mental effort.

251. Violation of this Principle.—Another illustration may be drawn from the frequent mal-practice in teaching which presents to the senses of the child pictures whose originals his imagination strives in vain to realize because it is not yet supplied with fitting material by the antecedent faculties. Take the vast objects that, in the study of geography, are represented on the maps by diminutive figures, signs, and contrasting colors that are wholly arbitrary. Now these merely conventional symbols presented to inexperienced eyes, suggest no concept of the imagination which answers to their colossal realities. Consider the countless observations, analyses and abstractions that must be accurately made and stored in memory, before the pupil can construct an image that corresponds with any fullness to the word "London" which is represented on the map by a solitary star. How is it possible to realize adequately the great rivers from contemplating zigzag lines of ink without having scanned some of the actual streams they symbolize? Did any youthful learner ever depict in his imagination the Alps or the Appenines from the trifling characters that bear their names? These are instances of vast aggregates which this study, by merely conventional signs, crowds upon the imagination[2] before it has either the training

---

[1] The study of the concrete must precede that of the abstract.
[2] Chap. XVI. 211.

or the materials which are indispensable to the forming of corresponding images that give vitality and value to the studies pursued in school. It is their constant and fruitless appeals to the undeveloped imagination of the juvenile student, that makes the study of geography the mere acquisition of names, which being associated with concepts that are to the last degree vague and dim, soon fade out from the memory and leave no trace behind.

The immense extent and complexity of its objects, the merely local association under which they are presented to the memory and the lack, in the juvenile imagination, of both strength and materials for making the corresponding images, demand, as seems to me, that geography should be deferred until the learner has more years, a greater power of conception, and a specific and thorough preparation.

Examples of similar mal-application of educational processes, may be found in the practice of assigning to primary pupils, those studies which present classified facts too wide for the grasp of the classifying faculty in its earlier stages. The science of English grammar as defining the parts of speech and developing the relations of these in the sentence, is frequently pursued in the most superficial manner, because its classifications transcend the antecedent experience of the young student. His classifying faculty[1] has not yet constructed the groups on which the definitions are based.

The definitions in grammar throughout demand a familiar knowledge of subtle classifications which are beyond the attainment of a mere child. In the example "an adjunct is a word, phrase or sentence used to limit a word" it is possible indeed, for a young pupil to learn

---

[1] Chap. XVI. 211.

and to recognize a word, a phrase or a sentence by its *form* and so to reach an appreciation of the genus involved in the definition. But it transcends his capacity to have analysed, compared and classified the subtle, metaphysical elements which constitute the species *used to limit a word.*

Grammar is in fact a subjective science and consequently, as we shall see farther on, is one of the studies that in a logical order, stands later in the series than the objective sciences which present their objects to the senses.

252. Untimely Effort a Serious Obstacle.—The plans of primary instruction that evoke the untimely effort of any faculty by presenting objects which are beyond its grasp, are serious if not fatal obstacles to early progress. The premature study of abstract numbers, as a further example, not unfrequently develops a repugnance to mathematics that lasts through life. The concepts of pure numbers are acquired primarily by instinctive analysis and comparison of concrete objects perceived by the senses. Infants and savages have only dim and rudimentary notions of pure numbers, because the processes of concrete conception, analysis and abstraction from which they are derived are, in the infant and the savage mind, rudimentary also. The course of instruction then that demands of the faculty of abstraction, the conception of these abstract ideas, subjects it to an unhealthy strain from which it naturally recoils. It were a far wiser policy to commence and continue the exercises on concrete objects (marbles, numerical frames and the like) by which the notions of pure numbers are gradually and regularly evolved through analysis and abstraction, until they are realized with the utmost ease and distinctness. In such studies the concrete should not only precede the abstract but should precede it with great circumspection and with an abundance of minute and accurate details. Nature's methods in the processes of mental unfolding, should not be simply followed, but they should be carefully systematized and improved.

253. Conventional Forms when Learned.—The teaching of conventional forms before the eye of the pupil is sufficiently trained by practice on forms that are more attractive because more simple and natural, is another violation of the principle that heads our chapter.

To the eye of a tyro, the arbitrary signs that compose the alphabet, are meaningless and often repulsive. They furnish none of the elements that arouse the instinctive spontaneities. The learning of the alphabet ought for this reason to be deferred until the visual attention has been adequately drilled and strengthened by formal practice on concrete forms that possess the characteristics of simplicity and beauty.

254. What the Serial Unfolding Demands.—The instances cited present a few of the many obstacles to juvenile progress arising from an unwise arrangement of the things to be studied. The law based upon the serial unfolding of the faculties demand:

1. That no advanced faculty in the intellectual series[1] should be stimulated by formal instruction until the preceding faculties have been trained to elaborate the precise materials out of which it can form clear and full concepts of the objects presented.

2. That each faculty, when called into exercise by formal instruction in its normal order, should be trained to a capacity for strenuous effort by the study of its own peculiar object in a series which commences

---
[1] Chap. XVI. 211.

with the simplest examples and closes with the most complex.

3. Studies whose objects, as explained above, are beyond the conceptive power of the faculty to which they appeal, should be deferred until the antecedent faculties have been adequately disciplined.

4. It is clear that those studies which present the objects to the senses and thus furnish in regular succession the materials which call each subsequent faculty into strenuous action, conform completely to the laws of mental growth and are, consequently most available in early education.

QUESTIONS ON CHAPTER XIX.

What are the injurious effects of an unwise choice or a faulty arrangement of studies for the young mind? On what does the period when a faculty is naturally active depend? Illustrate by examples. Malpractice in teaching by presenting to the imagination pictures whose originals it cannot realize. Injurious effect on the mind of studying geography without adequate preparation. Reasons why the study of geography should come later in the course. What are the reasons why English grammar is often studied superficially? Why cannot a child understand the classifications of grammar? The deleterious effect of untimely effort in the study of abstract numbers. Necessity of commencing the study of numbers with exercises on concrete objects. Why should not the child begin his first formal study by learning the alphabet. What does the arranging of studies based upon the serial unfolding of the faculties demand?

Chapter XX.

STUDIES MUST BE SELECTED THAT WILL DISCIPLINE THE FACULTIES STRICTLY IN THE ORDER OF THEIR DEVELOPMENT.

255. The Sciences as Gymnastics.—The criteria for determining the value of any science as the special gymnastic of a given faculty or group of faculties, are as follows:

1. The subject-matter of such sciences must supply the peculiar objects of the faculty it calls into exercise.

2. The objects presented by such science must be replete with the constant qualities that incite the faculty under discipline to efforts that are strenuous and effective.

3. The operation by which such science modifies or combines its subject-matter into other or higher forms, must be identical with that by which the faculty under discipline transforms its objects into its products.

4. This operation must be so exact, systematic and exhaustive throughout, as to demand from the faculty under discipline, efforts that are strenuous, persistent and effective.

5. When, as in most cases, a science incites to action a group of consecutive faculties in the series,[1] it must in fulfilment of the foregoing conditions, present to each in succession objects that are complete and operations that are logically perfect.

---

[1] Chap. XIII. 173.

256. Illustrated by Systematic Botany.—The above tests of the fitness of any branch of learning to serve as an intellectual gymnastic, may be illustrated in detail by a single example, systematic botany, a concrete science, presents objects, usually flowers, to the eye, that conform to the tests 1. and 2. The attention of the student must fasten upon these with a strenuous effort that, through the aid of memory, produces a concept that is replete with the constant qualities required for the perfect classification of science, this answers to the tests 3. and 4.

Again, the faculty of conception holds this complete and clean-cut concept of a flower vividly up while analysis notes with precision and stores in memory, each of its characteristic organs and properties, thus gaining and presenting, as an object for classification, the perfect concept of the flower and all its essential parts in vivid distinctness.

Finally, the faculty of classification acting upon the analyzed concept thus presented, proceeds, by comparing its parts and properties with those of other like concepts previously classified, to adjust it in its appropriate classes, namely, its variety, species, genus, etc.

Here we have in concrete botany a series of facts and operations which are completely identical with the succession of objects and efforts in the first division of the intellectual series. The science presents in succession the concrete flower, the minute description of its organs and properties and finally its appropriate classification according to these, and the processes employed call into strenuous action, consecutively and without a break, the powers of sense perception, memory, conception, analysis and classification.

From this perfect coincidence of object and operation, without and within, we may safely infer the gymnastic value of systematic botany in the elementary course of scientific studies. In fact, it fulfils completely the conditions set forth in test No. 5, which requires that a science which appeals " to a group of consecutive faculties shall present to each in succession, objects that are complete and operations that are logically perfect."

257. Zoology Submitted to the Same Test.— Concrete zoology addresses itself to the same group of faculties as that to which botany appeals. For example, this science designates and presents to the senses, first, the concrete animal with its general qualities of shape, size, weight, color, etc., secondly reveals by analysis its organs, noting in each the form, color, compactness, purpose and finally comparing these organs of the animal under scrutiny, with similar organs of other animals, it adjusts such animal in the proper classes. Thus elementary zoology is to be set down also as one of the earlier studies which may serve as a gymnastic for the faculties from sense perception to classification inclusive.

But zoology and botany though exercising as studies the same group of faculties, differ from each other in one particular. The objects of which zoology treats are more complex than those of botany and, consequently, demand severer efforts of attention in scrutinizing, analyzing and classifying them, zoology ought, therefore, to stand one remove later in the series of elementary sciences which are adapted to the training of the young.

258. Mineralogy also.— A third science, which deals with external objects and groups them in classes by identical processes as required by test 5, is that of mineralogy. This science treats of the forms and structure of minerals and the parts and properties by which they are classified. Its objects, however, are not so beautiful and striking nor are their qualities so distinct as those of botany and zoology. They may nevertheless be profitably employed as adding variety to the early series of studies and giving the pupil a knowledge of the things he constantly sees.

Since each of the above trio of sciences has for its purpose the arrangement of its exclusive objects into scientific classes, they are fitly called the sciences of classification. The educational value of these sciences, especially as means of training the faculties most closely allied to sense perception, has never been fully appreciated. Their claims to high distinction as early gymnastics, may be summed up as follows:

1. When taught with the aid of concrete specimens presented to the eye, they furnish all the incentives to strenuous, persistent attention.

2. Their objects may be so arranged as to form progressive series which beginning with the simplest examples, gradually rise through years of study to the more complex.

3. They impart to the pupil a knowledge of the objects around him and specially the forms of life with which he comes daily in contact, and may, for this reason, be reckoned among the useful branches of learning.

4. They beget in the pupil a love of natural beauty, bring him into active sympathy with all God's creatures, and thus exert a salutary influence in the moulding of character.

5. Presenting their concrete objects to the eye for the purpose of classifying them, they are, *as gymnastics*, the natural antecedents of physics and chemistry. For the subject matter of these sciences consists in the identical classes which systematic botany, zoology and

mineralogy have supplied. Logically, then the former begin where the latter end.

259. The Mathematical Sciences.—Arithmetic —Its Gymnastic Value.—The points in which arithmetic conforms to the standards for determining an effective gymnastic, can be easily and clearly stated.

In the first place, arithmetic, as the science of abstract numbers, deals with objects that are identical with the early concepts of abstraction. It therefore opens by presenting to this faculty a portion of its own peculiar products which are abstract notions of color, number, form, motion, etc., etc.[1] But notions of numbers realized primarily as abstract concepts, immediately become the objects of the classifying power by which they are easily wrought into the simple and perfect classes based on the one characteristic of *identity of units.*[2]

Again the facile and faultless classes of abstract numbers which arithmetic thus produces, evidently become the inexhaustible material on which class judgment and reasoning perform successively the unerring operations that reveal with absolute certainty the countless properties of numbers.

It is manifest from these facts that aritnmetic, beginning as it does with an appeal to abstraction, the fourth *conscious* faculty of the series, can not be classed along with the concrete studies that are adapted to the first juvenile training. Its true place is in the second series of primary studies where it is preceded by a special preparation in concrete numbers now generally given in the best primary schools.

But though arithmetic belongs to the second series of primary studies, it manifestly ought to lead this

---

[1] Chap. VIII.   [2] Chap. X. 148, 149.

series in the order of time. For both its subject-matter and its operations are identical with the objects and movements of the earliest abstract thinking. Probably, the first abstract notions gathered by the child are those of number. And these notions, limited and vague as they are, he at once puts together in simple groups that form rudimentary classes. On such classes judgment and reasoning operate with fitful movements that merge by degrees into conscious efforts, efforts which the science of numbers, with its simple and perfect processes, renders definite, precise and effective.

The chief characteristics of arithmetic may be summed up as follows:

1. The objects of which it treats, are absolutely simple—they display the pure, unvarying identity of numbers.

2. Its operations are, as a whole, the simplest examples of perfect deductive reasoning.

3. Among the sciences arithmetic stands conspicuous as the only instance of absolute independence. Its phenomena are those of number solely. Its processes are exclusively numerical. It borrows nothing outside of itself while its processes are indispensable as aids to the development of every other science. It is the monarch of the whole wide realm of the sciences.

4. Its operations are readily arranged in a consecutive order wherein each step depends on the steps preceding, and thus the entire science aptly illustrates the principle in education, that the operation of every disciplinary study should be adjusted as to advance gradually from the simple to the more complex.

5. Consequently the series of text-books on arithmetic are superior to those on other sciences and the instruction given by teachers in the same line, is, in general, correspondingly excellent.

260. Algebra.—Algebra deals fundamentally with numbers and is therefore essentially the form of advanced arithmetic. It differs however from common arithmetic in the following particulars—

1. Its numbers which are used as the measures of abstract quantity of whatever sort, are represented by the letters of the alphabet as symbols and the operations to be performed on these numbers, are indicated by the signs $+ - \times \div =$ etc.

2. The employment of symbols for its numbers and of signs for its operations, tends to obscure somewhat the real objects and processes it presents to the intellect of the student. Nevertheless, its objects are the abstract concepts of numbers used as units of measurement and its operations are the same in kind with those of arithmetic, though more abstruse and difficult.

3. In general the processes of arithmetic are employed in the transaction of business; those of algebra in developing the products of other sciences, especially of mechanics, astronomy and physics.

Beyond question, these facts show that algebra should follow arithmetic in the order of studies, especially as it constitutes the higher steps of a series that begins with simple numbers. But the student should be well versed in arithmetical processes, before he encounters a science which, though exercising the same faculties, is much more abstract in matter and method. When once engaged in it, he ought, as he proceeds, to give verbally the reasons for every step, the axiom on which it is based, and so to gain a complete insight into the logic of every process that leads to the final conclusion. Many students who are able to state the conditions of a problem with symbols and signs, go blindly through the *operations* they indicate, and so fail to secure the highest results sought in the study.

As a gymnastic, algebra manifestly stands prominent among the sciences that are of the highest rank as incentives to disciplinary efforts.

261. Geometry.—The sole phenomena which the science of geometry investigates, are those of abstract forms. It consequently presents its objects primarily to the faculty of abstraction, by which they are realized as its own peculiar products. The classifying act immediately follows, recognizes these abstract concepts of form as arranged with facility into different classes, which though less simple than those of number, are exact and systematic in the highest degree. The succeeding deductive processes by which the properties of each class are eliminated and proved, have all the characteristics of perfect reasoning. Hence it appeals to the same unbroken series of higher faculties and demands the same severe and systematic exertion of the reasoning power, as do the preceding sciences of number.

The distinctive features of geometry as compared with arithmetic and algebra are as follows:

1. The unvarying characteristics of the regular forms of which it treats, having a greater complexity than those of number, can not be grasped by the pupil at so early an age.

2. The classifications of geometry are made on the basis of resembling parts in each figure of a class. Those of arithmetic are made on the basis of identity in the units of each denomination.

3. Every link in the chain of geometrical reasoning is distinct and clear to the understanding, and is obviously essential to the validity of the demonstration.

4. Since the consecutive steps are not written down

by the student as he proceeds in a demonstration, the reasoning of geometry demands a more *sustained* effort of attention than that of arithmetic or algebra.

Thus in the clearness of the definitions that embody its matter; in the manifest connection of each step with the preceding and the following steps, and in the unceasing effort it requires from the first premise to the final conclusion, geometry is without a peer in the hierarchy of the sciences. As a gymnastic it fulfils completely the conditions required in No. 5.

As an elementary study, its place is alongside of, or perhaps a little in advance of algebra. But it must at any rate be preceded by a thorough knowledge of arithmetic.

262. Language:[1] **Names: Reading: Writing.**— The language series which is manifestly indispensable to a systematic education, is formed by a consecutive arrangement of studies that so obviously accords with the series of psychological powers as to need only a brief explanation. This series comprises:

1. *The names* which designate visible and tangible things, their acts and relations to each other. These names of objects and acts within range of the senses, as pronounced for the ear, uttered by the tongue and fixed in the memory of the learner, constitute the first lessons in language and call into early action the faculties of perception through the ear, of memory through association of sign with thing signified, and of conception as coupling together things and their names.

2. *Reading* or learning and pronouncing words in print which represent the audible words included in the preceding lessons and many additional ones. These printed words are addressed to the eye, pronounced

---

[1] Chap. XVII.

to the ear, uttered by the tongue, associated in memory by the triple connection of sign, sound and thing named thereby, and so re-produced by conception which unites in one vivid concept the thing and its name as addressed both to the ear and to the eye. Thus the exercises in reading employ effectively the first three faculties, sense-perception, memory and concrete conception. It is manifest further that, as the pupil advances, the higher examples presented for practice, will call analysis and abstraction into activity and that, especially when new words suggest objects which the senses of the reader have never encountered, they excite his imagination to an earnest endeavor to realize their meaning. This constant appeal to imagination in reading, suggests the necessity of a progressive series, wherein all new words shall denote objects or acts that are within the range of the pupil's experience. That is, every strange word should specify an object that belongs to a class some of whose members the reader has already observed and gained therefrom the materials for constructing an image concept that shall realize its meaning. Reading exercises the eye, the ear and the tongue and appeals constant to the imagination as the lessons advance; consequently, there is danger that such appeals shall be made without adequate preparation on the reader's part to answer them. Definitions do not avail if they comprise classes that he has never formed.

3. *Writing* gives practice to the eye and the hand and in connection with drawing, affords the means of effective discipline especially to the latter. It employs also increasingly the ear and the tongue, trains memory and conception by the exactness of the forms it presents, and, in its higher practice, appeals to imagination like the

exercise of reading. Consequently the same suggestions apply to both. Writing should be preceded by simple drawing lessons, should begin with writing the alphabet and be made the gradual means by which early reading and *sight* spelling is learned. It should finally attain, with scrupulous and prolonged practice, such facility as will justify its use in the early exercises of composition.

263. Composition.—Composition which employs subordinately the eye and the hand makes strong and incessant drafts on the memory through the effort of recalling ideas and their names, lends great vividness to such ideas when recalled, quickens analysis and abstraction, and excites the imagination to vigorous endeavor. Composition calls into independent action all these faculties *particularly* and all the rest of the *series subordinately*. It is a powerful instrument of intellectual discipline but needs, like all powerful agencies, the most delicate handling.

The practice of composition should commence and be carried forward in the following order:

1. Brief and simple written descriptions of objects within range of the senses.

2. Similar descriptions of objects that are familar but beyond the present reach of the senses.

3. Narratives, short and simple, of common events that are taking place near at hand.

4. Narratives of the most interesting events that the memory of the writer can furnish.

5. Short and simple papers on the most obvious duties of children, such as "truth telling," "obedience to parents," "kind treatment of our fellows," etc. Such articles should be preceded by brief remarks from the teacher illustrating each topic with fitting examples.

6. Topics that call out the first attempts at invention

and, therefore, require more exclusively sustained efforts of the imagination.

7. Along with this work in composition, should be given progressive and ample practice in the making of these elementary combinations in language that lead to the study of grammar. After models furnished by the teacher the learner should construct the simple sentence in each of its three forms, transitive, intransitive and neuter, learning to distinguish successively the parts as he makes them. Next let him repeat the sentence forms and modify the subject with an adjective; then construct again the same forms with other words and modify the verb with an adverb; then, with forms again renewed, modify the subject with an adjective phrase and the verb with an adverbial phrase, first learning thoroughly the parts of the phrase forms.

In this way, the forms and office of even the adjective and adverbial sentence, could be taught by first constructing and naming the form and then noting its effect on the word to which it is appended. These language forms should be familiarized by innumerable examples of the pupils own making, which ought to cover two or three terms of practice and constitute an adequate preparation for the more difficult study of English grammar. As they require the constant writing of words, they supercede at this stage of study, the necessity of formal spelling lessons.

This course in the construction and discrimination of language forms, employs systematically and effectively, the earlier faculties of the series from sense perception to imagination and classification inclusive, and incidentally all the other faculties.

264. English Grammar.—The science of grammar deals with the classified offices and relations of words, which it presents by means of definitions. The classes of words with which it begins, are therefore the products of the classifying faculty, and it is on this faculty that grammar makes its first draft. The analytical reasoning that follows and reveals the various relations of words, as the elements of the sentence in its different forms, exercises powerfully class judgment and the reasoning faculty—consequently English grammar, as a science, has its proper place among the advanced gymnastics, though many a school still assigns it as an elementary study despite of the fact that its results are uniformly valueless. If the writers of its text-books would accept and apply the following suggestions, the science of English grammar would supply all the requirements of an effective gymnastic.

1. The science is based upon the genius and structure of the English language exclusively, and not on those of Latin.

2. Each definition in English grammar should include rigidly every instance of the thing defined and exclude every other.

3. The rules in English grammar should rest on *general* facts and not on exceptional ones.

4. In English the inflections are few and consequently the relations of words to each other are indicated mainly by their office and position and not by their form or endings.

5. Grammatical analysis should in all cases reveal with precision the office and relation of each single word in whatever compound it may be found.

265. Rhetoric.—Rhetoric, as a science, deals also with classified concepts which it presents to the classifying faculty at the outset. Its matter comprises those systematized forces of language which render it, in the highest degree effective as a means of expression. Rhetoric inquires what properties of words and what forms of the sentence, will display most vividly the idea or emotion to be conveyed to another. In short, rhetoric reveals fully the aptitudes of language that enable the reader to grasp the thought with the least possible waste of effort expended on its symbol. To this end rhetoric signalizes the properties of words that render them most forcible, such as brevity, specificness, melody, novelty, suggestiveness of sound, etc., etc. It elucidates, moreover, the highest effectiveness of the sentence as gained through the various methods of arranging it. Moreover, this science shows exhaustively the fitness of metrical language to convey the beauty and the imagery of poetic thoughts.

Now all these operations consist largely in the comparisons of class judgment and the inferences derived therefrom respecting the particular elements or forms of language under examination: rhetoric is therefore a deductive science and as such serves as an effective gymnastic for the reasoning faculty. The classified facts which constitute its materials, are the products of previous induction. Like all other language sciences, it also draws constantly on the imagination and the faculty of analysis. Since the subject-matter of rhetoric is composed of the grammatical elements, grammar is its necessary antecedent in the order of studies. Aside from its gymnastic value, the knowledge which rhetoric imparts is indispensable to effective writing and speaking. Consequently along with the study of its principles, should go the frequent exercises by which they are reduced to practice.

266. Geography.—Let us consider briefly the facts which geography presents, the faculties to which they appeal and the intellectual operations by which they are wrought into the forms of knowledge.

First, the facts comprised in this branch of study are

threefold, namely, physical, political and astronomical.

1. The facts of physical geography are the vast and multitudinous objects that constitute the surface of the earth, together with the animals and plants that subsist thereon.

2. The facts of political geography are the men that people the earth's surface, their races, habits, manners, customs, character, government, and the divisions of land they occupy.

3. Astronomical geography classifies and explains the facts that arise from the relations of the earth as one of the planets in the solar system.

The immense objects embraced in physical geography, being beyond the range of the senses, are presented first by definitions in classes based on superficial resemblance and, secondly, by descriptions addressed, of course, to the imagination of the student. The extensive objects with which political geography deals, are substantially similar to the above both in superficial classification and in the image making power to which they appeal. To determine then the value of geography as *an early* gymnastic we have simply to answer the following questions:

1. Has the early imagination attained the discipline which enables it to respond effectively to the demand which the study of geography makes upon it? Has antecedent observation yet gathered the materials out of which imagination can form pictures that represent the vast originals designated by inadequate symbols on the map?

The truth is, the study of geography, even with the most careful illustrations in map drawing, requires previous preparation in the elements of botany, zoology, mineralogy, in drawing the natural objects, trees, buildings, roads and landscapes within visible range, and in personally observing, so far as possible, such objects as towns, cities, rivers, lakes, forests, plains, mountains, etc., etc. With these materials previously gathered, the pupil is able to construct the image concepts that constitute geographical knowledge. Lacking these, imagination responds feebly with pictures that are vague and false and the study of geography lapses into a mere memory of names which soon fade out forever.

Whatever its value as a gymnastic, the knowledge which this study bestows is of the highest importance, it is, in fact, indispensable to every well informed person; and consequently geography will maintain its place in our common schools. But coming later in the course and preceded by adequate preparation, the evils that attend it as a primary study, will disappear.

267.—Table showing the earlier Operations in concrete Numbers and the later Operations in Arithmetic that follow in the same Series.

Faculty.	Object.	Action.	Product.
SENSE PERCEPTION.	Things representing concrete numbers.	Observing things; specially as numbered, added or substracted, etc.	Percepts of things counted, added, etc.
MEMORY.	Percepts of things counted, added, etc., etc.	Acquiring, retaining, recalling concepts of things counted, added, etc., etc.	Unconscious concepts of things counted, added, etc., etc.
CONCEPTION.	Unconscious concepts of things counted, added, substracted, etc.	Representing things counted, added, etc., etc.	Conscious vivid concepts of things counted, added, etc., etc.

Faculty.	Object.	Action.	Product.
ANALYSIS.	Conscious vivid concepts of things counted, added, etc., etc.	Analyzing concepts of things counted into the numbers they represent.	Concepts of the particular numbers represented by the things analyzed.
ABSTRACTION.	Concepts of particular numbers represented by things analyzed.	Abstracting numbers from the things analyzed that represented them.	Abstract concepts of numbers gained from analysis. [*Concepts of abstract numbers presented by the science of arithmetic which begins here.*]
IMAGINATION.	Concepts of concrete numbers: abstract concepts of numbers supplied by abstraction and arithmetic.	Forming image concepts that represents numbers.	Image concepts that vaguely represent numbers.
CLASSIFICATION.	Concepts identical with those of imagination.	The facile classifying of abstract numbers from arithmetic.	Class concepts of abstract numbers.
CLASS JUDGMENT.	Class concepts of abstract numbers from arithmetic.	Affirming the relations of class concepts supplied by arithmetic.	Affirmed concepts of relations of the class concepts of arithmetic.
REASONING.	Affirmed concepts of relations of the class concepts of number.	Comparing these relations as affirmed by class judgment.	Deduced concepts of new truths reached through the perfect reasoning of abstract arithmetic.

268. Classification of the Sciences.

Sciences of classification
   (Subj.-matter: concrete forms)............... { Botany. Zoology. Mineralogy.

Sciences of perfect deduction
   (Subj.-matter: abstract numbers and forms).... { Arithmetic. Geometry. Algebra.

Sciences of expression
   (Subj.-matter: words)...................... { Grammar. Rhetoric. Criticism.

Sciences of induction
   (Subj.-matter: laws of matter)............... { Physics. Chemistry. Physiology.

269. Table Suggesting Partial Series of Studies.

Language Series.	Mathematical Series.	Nat. Science Series.
1. Names of things.	1. Concrete numbers and operations.	1. Lesson on colors.
2. Drawing and learning the alphabet.	2. Concrete regular forms and drawing.	2. Lessons on parts of organic forms with drawing.
3. Elements of writing and spelling.	3. Simplest abstract numbers, fundamental operations.	3. Lessons on plant forms and drawing.
4. Writing.	4. Arithmetic, common.	4. Lessons on animal forms and drawing.
5. Composition.	5. El. geometry.	5. El. systematic botany.
6. El. grammar.	6. El. algebra.	6. El. Zoology.
7. El. rhetoric.	7. Arithmetic, ad.	7. Geography.
8. Ad. composition.	8. Geometry, ad.	8. Mineralogy and history.
9. Ad. grammar, or foreign language.	9. Algebra, ad.	9. Structural botany and zoology.
10. Ad. rhetoric, or foreign tongue.	10. Trigonometry.	10. Physics.
11. Criticism and English literature.	11. Descriptive or analytical geometry.	11. Chemistry and physiology.

QUESTIONS ON CHAPTER XX.

Give the five criteria for determining the value of any study as a special gymnastic. Show how these are illustrated in the study of early systematic botany. What place then should elementary botany hold in the series of early studies? What faculties does elementary zoology call into exercise? What place should it consequently hold in the series? Possible value of elementary mineralogy as an early gymnastic? Sum up the claims of the three preceding studies as early gymnastics. What is the character of the objects with which arithmetic as an abstract science deals? What is the nature of the classifications in arithmetic? To what faculty does the science of arithmetic present its objects? Describe the rudimentary operations of the young mind in numbers. What is the character of the reasoning processes which the study of arithmetic requires? Sum up the chief characteristics of arithmetic.

With what entities does algebra fundamentally deal? In what three particulars does it differ from common arithmetic? What are the conditions on which algebra may follow arithmetic in the order of studies? Value of algebra as a gymnastic. What are the sole phenomena which geometry investigates? To what faculty does it primarily present its objects? What is the character of the classifications in geometry? Value of the exercises that geometry supplies to the reasoning faculty. Give the two distinctive features of geometry as compared with arithmetic and algebra. Nature of the demand which the study of geometry makes on the attention of the student. What is the comparative value of geometry as a gymnastic? What does the early language series comprise? The learning of names constitutes the first lesson in language. Effect on the early faculties of learning to read. Describe the successive steps of the process, and name the faculties to which they appeal. What early faculties does writing call into exercise? By what simple lessons should writing be preceded? What faculties does composition primarily exercise, and what is its value as a gymnastic? Give the progressive steps by which composition should be commenced and carried forward. Show how composition may be used to introduce the study of elementary grammar, and what faculties are exercised thereby. What are the objects with which the science of grammar deals? Upon what faculty does grammar make its first draft? What faculties does this science most powerfully exercise? What place should grammar hold in a series of studies? Give the characteristics of grammar as a science. What does the material with which rhetoric deals comprise? What are the various steps this science traverses? Of what faculties is rhetoric an effective gymnastic? What is the value of the knowledge it gives us? What are the three branches into which geography is divided? What facts does physical geography present? What are the facts which political geography embraces? What facts does astronomical geography explain? Is the study of geography suitable as one of the earliest gymnastics? Give the reasons that demand an answer in the negative. What previous preparation does the study of geography require of the pupil? To what faculty mainly does geography present its objects? What is the value of the knowledge which geography supplies when rightly studied? Give the table showing the earliest operations of concrete numbers and the faculties they call into exercise. What are the sciences of classification? Of perfect deduction? Of expression? Of induction? Write out three consecutive series of simultaneous studies, which are based on the foregoing facts and principles.

Chapter XXI.

ARRANGEMENT OF STUDIES AND METHODS OF INSTRUCTION IN EARLY EDUCATION.

270. The Period When Formal Education Should Begin.—The formal education of the child should begin whenever he is capable of applying the intellectual senses to their objects by an act of attention.[1] This power of attending to a single thing exclusively, though still fitful and rudimentary, may now be directed to its appropriate objects and moderately exercised. But a judicious teacher will not fail to recognize the feeble character of the voluntary efforts which the tyro is able to put forth. The power of centering upon an object presented to the senses, is, as yet, only a minor element in the spontaneity that still largely prevails. The sense appealed to needs at this period to be gently incited to effort by presenting objects that are naturally attractive and by removing all distracting influences.

The supreme danger at this incipient stage of formal instruction, is that the exercises shall take the character of tasks[2] which become permanently repulsive. The child naturally hates whatever is wearisome, specially in the way of mental exercise. The tender organ cannot bear a protracted strain; only the brief effort which the simple beauty of the object naturally induces, is healthful and strengthening. All overstrain

---

[1] Chap. I. 10.   [2] Chap. XIX.

is weakening and often fatal to future progress. The first formal lessons ought, therefore, to be frequent and short, rather than few and prolonged. It were wiser to close the exercise while the interest of the pupil is increasing or has reached its acme, than to continue it until the interest is on the wane. For the natural avidity with which sense-perception fastens upon its fitting objects, is, at this juncture, the only true stimulant to effort, and the sole question is how to sustain and intensify this interest from day to day and from year to year. The practice of giving to the little ones intellectual tasks as a punishment for disobedience, or of making their sessions at school equal to that of older pupils is, let us hope, a barbarity of the past.

271. The Age When Incipient Systematic Efforts Begin.—The age at which to commence systematic training of the senses by persistent exercise upon their objects, while varying in different children, will, on the average, be between the fifth and sixth years. Up to this time, the infant mind has been under the dominion of the spontaneities and the intellectual senses. Along with the faculties they incite, have acted mainly in fitful response to the solicitation of their objects, without interference of the will. These years of infancy that precede the age of formal effort, may be fittingly named the play period; for play is the spontaneous activity of the faculties excited simply by the presence of their objects.

272. Value of Play Schools.[1]—But even the spontaneities of infancy need not run to waste. Advanced educators have unanimously recognized their signifi-

---

[1] See Payne on Froebel's Kindergarten.

cance and assigned them a prominent place in the system of complete education they advocate.

The play schools of kindergartens of Germany, now conducted also in the large cities and large towns of England and this country, are the evidence that the world is beginning to appreciate the advantages of guiding aright the instincts and spontaneities of the infant. It is not the purpose of the play school to elicit or encourage special voluntary effort. It deals with the infant spontaneities solely as spontaneities and seeks to lead them to the truest action by the allurement of objects and plays wisely selected and skilfully arranged. By this happy expedient it quickens and aids the processes of nature, awakens a love for the simplest forms of beauty and novelty, begets the habit of quick response to the presence of the most approved objects, and prepares the infant intellectually and morally for the stage of formal instruction that is to follow. The little ones are fortunate indeed who enjoy either at home or in the kindergarten, the advantages of a genuine play school. The mother is nature's first teacher and a salutary effect on the race would surely follow if she had universally the skill and the leisure to be the sole conductor of the play school. For the ideal home is a play school where the sports of infancy lead perpetually the dawning intellect and tender heart to what is beautiful and good, and the too frequent defects of the real home are the only reasonable grounds for the existence of an outside kindergarten.

273. Objects for the Formal Training of the Hand.[1]—Under normal conditions, the objects employed for the training of perceptive touch *exclusively* are

---

[1] Chap. III. 70, 71.

limited in number. In nearly all the manual exercises whose purpose is to give quickness and delicacy to the perceptions of touch, the eye participates not only in the operation but in the discipline that results. The only separate practice given to the hand occupies the interval of earliest infancy before it has taught the eye that the distinctions of light and shade, mean solidity and shape. Thereafter when the eye has learned from the suggestions of the hand to discern form and distance through the variations of color, it far outstrips the tardier movements of its teacher and takes the principal part in the joint discrimination of properties that appeal to both. The only condition in which the hand is trained separately to the greatest range and delicacy of perception of which it is capable, is that of blindness.

This exclusive training of touch to perform successfully the functions of sight, while it demonstrates the expertness which special education can give to the faculty, would of course be a waste of time to those who are blessed with eyes that see. Under normal conditions, the three intellectual senses, touch, sight and hearing, whose triple function is to reveal the phenomena of the external world, should receive a simultaneous training[1] that will fit them most effectively to act in harmony as feeders of the intellect.

274. Objects that Train the Sense of Sight Exclusively.[2]—The immediate objects of sight, be it remembered, are the diversities of color, light, and shade. Subsequently, as we already know, under the early tuition[1] of the hand, the infant eye cognizes solidity and shape through light and shade which are their subtle signs. Manifestly then since the colors are the

---

[1] Chap. III. 71, 72.   [2] Chap. III. 71.

earliest and the only *immediate* objects of sight, a series of lessons in which the varieties of color are discriminated and named, will naturally be first in the order of time. Moreover because modifications of color are the signs to the eye of the countless modifications of form, a visual perception of the one depends wholly on a knowledge of the other. "First the sign and then the thing signified," is the only true order of acquisition in all cases wherein the thing can be learned only through its sign.

274. The Natural Order of Presenting Colors to the Eye.—The primary, secondary and tertiary colors.[1] The child will undoubtedly come to his first formal lessons at school with some knowledge of colors gained either at home or in the kindergarten. The purpose now, however, is to acquire and strengthen the habit of attention through sight, by directing the eye steadily to its simplest objects. The color series to be used for this end, whether composed of pigments or natural objects, or both, should include all the fundamental colors, simple and compound, with their numerous varieties of tinge, tint and shade. Here, as elsewhere in all educational exercises, the lessons should begin with the simplest and most vivid examples and progress, step by step, on to the most complex and less vivid ones.*

The pure primary colors, red, yellow and blue, being the simplest and most vivid of all colors, fulfil the conditions required in the objects presented to the eye for its first regular exercise.

---

[1] Chap. XVII. 211.

* The most convenient equipment for giving the lessons on colors, are the color charts, but the enthusiastic teacher will add to these collections of natural objects such as flowers, leaves, feathers, woods, minerals—whereby a wide range of hues may be illustrated.

Let the pupil scan and name each of these and repeatedly select it from miscellaneous groups of colors displayed by various objects before him. Repeat the exercise until he can instantly recognize and name red, yellow, or blue, in whatever form it meets his eye. Proceed next to the pure secondary colors, green, purple and orange and, in a similar way, impress each one on the attention of the pupil exclusively, until both the name and the notion of it are firmly fixed in his memory. Then unstinted practice in elementary analysis may be furnished by requiring him to distinguish every specimen of green, purple or orange when mingled promiscuously with other colors.

Our next step in the carefully conducted series, will deal with the tertiary colors, russet, olive and citron. Here we employ the same processes with scrupulous exactness as follows:

1. The separate color inspected and named on the chart again and again.

2. The scrutiny of many specimens of it presented in different natural objects.

3. The unaided selection by the pupil of the same color from a multitude of things that are differently colored.

All these steps must be reiterated until they reach, in the mind of the learner the perfect familiarity of spontaneities.[1]

276. The Tints and Shades.—Now return to the primary colors and continue the lessons in "sight-training," presenting in succession exhaustive examples of the lights and shades. Each one of the primary, secondary and tertiary colors may be prepared for the lessons now in hand, by tinting it and thus producing

---

[1] Chap. XVIII.

successive specimens that finally end in pure white. Successive shades of the same color may also be made by mixing it with increasing quantities of dark pigment until the series ends in black.

Beginning with red, direct the attention of the pupil to, say, six specimens which illustrate its tints as progressing from pure red to white. Let him scrutinize and name each specimen in succession as "first tint of red;" "second tint;" pink, etc., etc. Let him next examine identical tints as found otherwise than on the chart, and then, in a collection of miscellaneous tints, identify by comparison those which he had previously noted in the separate series. The same course of exercises may now be given in the shades of red, running with six samples from red to black.

Following in gradual succession come the tints and shades of yellow and of blue, each of which should be scrutinized as above and called by its specific name, if it has one, or otherwise designated by the ordinals 1st, 2nd., 3rd., etc. The tints and shades of green, purple and orange, treated in the same manner, come next in order and finally the tints and shades of russet, olive, citron, close these primary exercises on the modifications of the regular colors called their tints.

277. The Tinges.—Having dealt with the tints and shades so thoroughly that the juvenile eye can thereafter detect and name any single instance of them at a glance, we may now proceed to the practice of sight-efforts on the tinges.

A tinge is the modification of any color produced by adding to it a limited portion of red, yellow or blue. For instance, if we add to pure red a moiety of blue we have *crimson* or red tinged with blue. Increased quantities of blue will produce stronger and stronger tinges until we reach purple, which is one of the secondary colors. On the other hand, if we mix with pure red a limited quantity of yellow, the result will be *scarlet*, or red tinged with yellow. Augment the yellow by degrees in several specimens, and you have red with tinges of yellow which increase in brightness until you reach orange, which is another of the secondary colors. In the same way, pure blue may be tinged with red, by degrees, on the one side, until we arrive at purple; or, on the other, it may be tinged with yellow until we reach orange. Or pure yellow may be mixed with a small portion of red which, increased uniformly in successive steps, yields a series of yellows tinged with red that closes with orange; or, on the other side, yellow may be tinged with blue which, with added amounts at each remove, advances to pure green.

By similar steps, as the competent teacher well knows, we obtain for the eye of the pupil, the tinges of each of the secondary colors, green, orange, and purple, which end in corresponding tertiary colors. An example or two will suffice. Green is a compound of yellow and blue in definite proportions. Now produce examples of green tinged with red, by adding the latter in increased quantity for each example, and you will finally obtain russet which is one of the tertiary colors composed of blue, yellow and red the latter predominating in the mixture. Similarly orange, a compound of red and yellow, may be tinged with blue successively for samples, until the final result is olive, a tertiary color wherein blue predominates; or purple, composed of blue and red, may be repeatedly blended with yellows, producing a series of purples with yellow tinges that culminate in citron, a tertiary color in which yellow is the prevailing element.

The above methods of producing modifications of the pure colors by tinting with white, shading with black, or tinging with red, yellow or blue, are presented to the teacher, for the sole purpose of enabling him to arrange this primary series of color lessons in such order that it may, in all its parts, proceed from the simpler to the more complex.

The tinges may be used for eliciting the early discriminating efforts of sight, with the same consecutive acts of attention as suggested for the lessons previously given on the pure colors, their compounds, and their tints and shades. That is, the learner: 1. Scrutinizes and names the single sample of a tinge on the chart; 2. Recognizes the same sample as exhibited in other objects; 3. Selects it from miscellaneous groups of different colors in which it is found. Finally, these color lessons which should be protracted by the necessity of giving complete familiarity with each step before proceeding to the next, may close with the inspection of the irregular tertiary colors; namely, the browns, the grays, the drabs, etc.

Of course it is understood that, in these elementary exercises on colors and their variations, no attempt should be made to teach the child the means of producing them. His mental efforts are properly limited to the acts of perception directed repeatedly to single successive examples; and under persistent practice rightly guided in this line, the power of discrimination that results will surprise the teacher; for no teaching is so successful as that which follows the order of nature. The combination of colors; the production of harmony by the contrast of complementaries; the philosophy of scenic effects and the scientific theory of colors as derived from the spectrum, will constitute one of the profitable studies when the higher faculties come into play. They may, consequently, be postponed to that period.

For convenient reference by the teacher, I append the following formulæ, which embody the artist theory for making the compound colors with pigments that are pure. The figures used as coëfficients, indicate parts of pure colors which are equal *in weight*.

278. Formulæ for Producing the Compound Colors.—3 Yellow + 5 Red + 8 Blue = White (standard of complementary colors.)

5 Red   + 8 Blue = Purple. ⎫
3 Yellow + 8 Blue = Green. ⎬ Secondary Colors.
3 Yellow + 5 Red = Orange. ⎭

3 Yellow + 5 Red + 16 Blue = Olive. ⎫
3 Yellow + 10 Red + 8 Blue = Russet. ⎬ Tertiary Colors.
6 Yellow + 5 Red + 8 Blue = Citron. ⎭

279. Objects used for the Simultaneous Training of the Hand and Eye.—Along with the color lessons, which appeal to the eye alone, may be given a series of lessons in elementary forms, which are suitable for the incipient training of the eye and the hand, in concept. Before proceeding to the arrangement of forms most available for this purpose, let us settle fully and finally, the criteria by which the objects presented for the first formal exercise of the intellectual senses, should be selected and arranged:

1. They should be of such size as to come within easy range of the senses whose action they evoke.
2. They should be the simplest examples that can be devised or selected, that is, out of the multitude of complex objects which nature displays simultaneously to the senses of the child, those forms only are suitable which present the least number of characteristic outlines. The regular figures are, therefore, best adapted to the work in hand.
3. They should contain, so far as the case admits, the simplest elements of beauty and novelty.

4. They ought, in every feasible instance, to be of such a character that the pupil under instruction, can make imitations or copies of them.

5. Beginning with the simplest regular forms, they should progress by slow and punctilious efforts to the more complex, and finally to the irregular ones.

With these criteria clearly in mind, let us attempt the arrangement of a series (or several series) which shall approach the nearest conformity with them, that is attainable in the nature of things.

280. The Forms of the Triangle.—Of the solids and the plane figures into which all regular forms are divided, the plane figures containing the fewest outlines, are the simplest. Of the regular plane figures, those which present the fewest characteristic outlines, are the triangles; and consequently the forms of the triangle will furnish the material for our incipient lessons on form.

Bear in mind, to begin with, that we are not giving lessons in abstract geometry. Having as yet no knowledge of classified truths in this science, the child cannot appreciate the definitions that are based upon them. We cannot, therefore, profitably define the triangle in our early form lessons, even granting that the definitions are of the simplest. Our exclusive purpose is 1. To impress an elementary form on the eye of the pupil; 2. To incite his tongue to pronounce its name heard from our lips; 3. To encourage and direct his hand in repeated endeavors to copy it with the pencil or the crayon, and 4. To repeat these rudimentary processes, until the pupil goes through them with spontaneous readiness.

281. Equipment.—A word here as to the simple equipment needed for giving the primary lessons on form. Painted tin, cut neatly into plain geometrical figures, including all the varieties of single, regular, superficial forms found in the ordinary text-book, will constitute the principal feature of the proper outfit. Ordinary drawing paper, a graduated ruler, a pencil and pair of scissors for each pupil will complete the collection for the present. A pair of compasses may afterwards be added.

282. Manual and Visual Exercises on Triangular Forms.—The various triangles will, as we have said, supply the objects for the first series of simple form lessons appealing to the eye, and the hand, and exercising indirectly the ear and the tongue. The following order may be accepted as a model. Take, as the simplest sample of the form in hand, the tri angle which has equal sides and equal angles, display the concrete specimen to the eyes of the glass and let them repeat the single name you give it. Explain in the simplest words at command, that the flat side is its surface and show that it has two such surfaces. Then the class uniformly repeating in concert the terms you use, point to its edges, say that they are the "sides of the triangles." Let the pupils count and call them "the *three* sides of the triangle." Show by applying the ruler's edge that they are straight and by actual measurement that they are equal in length; and let the class repeat " this triangle has three straight sides that are equal in length. It is for that reason an equilateral triangle." Turning next to the angles point out how each one is a point or corner made by the meeting of two straight lines or sides. That the triangle has by count three angles. Copy one of these angles accurately on the board and fit each of the other two angles to the copy, thus showing

---

[1] Chap. XVI. 211.

by their exact coincidence that the three angles are equal. The class following each step with eye and tongue, repeat together " The three angles of this triangle are equal. It is therefore an equiangular triangle.'

Proceeding next to an exercise in elementary drawing, make on the board with crayon, three points so located that each is equi-distant (say six inches) from the other two. Instruct each of the class to make points similarly placed on board or paper. Then guiding the crayon by the edge of the ruler, connect the points with straight lines. Let the pupils carefully imitate the operation and so complete their first drawing. Let them repeat the process accurately half a dozen times and then, dropping the ruler, try to connect the points with straight lines, guiding the hand with the eye simply. Continue this incipient free-hand practice until the tyro can draw an equilateral triangle with a degree of facility and correctness.

The next lesson conducted in similar steps, will consist perhaps in scrutinizing, naming and drawing the parts of a right angled isosceles triangle or a triangle containing a right angle formed by the meeting of two sides which are of equal length. Evidently it is necessary that the pupil, before engaging in this exercise, should gain from particular examples, the precise notion of a right angle.

Draw a cord over the surface of quiet water and it is a horizontal line. A rope stretched along the level surface of a garden, represents a horizontal line. A railroad track running on a level grade is likewise a sample of a line that is horizontal. Now tracing a horizontal line on the board, thus, if I draw another line upward from its centre in such a direction that it does not lean towards the horizontal line on one side or the other, it is a perpendicular line. If it leans towards the horizontal line either way then it is not a perpendicular line. Now this line, if perpendicular, meets the horizontal in such a way as to form two equal angles—one on each side of it, these equal angles so formed are right angles. Incline the perpendicular towards the horizontal line on the right hand and it forms an acute angle which is less than a right angle, while on the left it makes an obtuse angle which is greater than a right angle.

The distinctions between right, obtuse and acute angles being clearly perceived as displayed in the samples provided, proceed to an exercise on the right angled isosceles triangle, the pupils first scanning, comparing and naming the parts, as suggested in the previous lessons. Then make a point on the board where converging lines shall form your right angle and locate to the right, say, twelve inches from the first, a second point which shall serve for the terminus of a horizontal line and twelve inches directly above the first, a third point for the terminus of a perpendicular line. Draw straight lines connecting each point to the other two and let each of the class leisurely make several copies noting and naming, as he proceeds, the horizontal, the perpendicular sides and the hypothenuse. The two acute angles may be shown, by coinciding measurements, to be equal to each other and together to equal the right angle.

The above samples of lessons that employ the eye and the hand with the simplest forms, may serve as guides in further exercises on the irregular triangles, which present examples of unequal sides forming, by their junction, obtuse and acute angles of different sizes.

283. Other Plane Figures.—From the triangles the lessons may advance to the higher plane figures and follow their grades of growing complexity which finally end in the circle. The square, the parallelogram, the pentagon, the hexagon, the heptagon and all the polygons, that follow farther on, may each be made the means of calling forth a series of progressive exertions through the eye and the hand.

284. The First Square and Parallelograms.— The first square may be produced by assuming the hypothenuse of a rectangular isosceles triangle as the hypothenuse of a similar equal triangle drawn opposite the first. The common hypothenuse of the two rectangular isosceles triangles thus drawn, becomes the diagonal of a square formed by their juxtaposition.

The first sample of an oblong parallelogram may be made by drawing through the square a line which shall be perpendicular to two sides that are opposite each other and parallel to the other two.

285. Added Suggestions.—Let the above exercises be conducted with careful circumspection; with many repetitions;[1] and with scrupulous attention to each step, inciting the pupil to the highest excellence both n the *knowing* and the *doing* which it requires. Be in no haste to get over the ground; the day of sacrificing accuracy and readiness for extent of knowledge, is past. Above all, stimulate the learner to make, so far as possible, *unaided* exertion. It is yours to call out and guide his efforts, not to substitute your own for them. Finally, after the pupil has reached satisfactory facility and readiness in delineating each figure, let him trace on stiff paper and, with a pair of scissors, cut out a sample of it for himself. He will, in this way, gain a complete equipment of his own producing, in which he will take no little pride.

---

[1] Chap. XV. 209.

286. The Squares Drawn on the Three Sides of a Right-Angled Triangle and the Circle.—The series of exercises for training the eye and the hand by means of the mathematical plane figures, may fittingly close with scanning, naming and drawing successively the parts of three squares described on the sides of a right-angled triangle; and then proceeding, with a pair of compasses in the hands of each pupil, to draw a circle and to learn therefrom its divisions and properties. The first may serve as an appropriate review of the preceding lessons on triangles and squares, with the added advantage of increased complexity in the figure to be copied. Proceed as before, locating the points, connecting them with straight lines, first guided by the ruler's edge and afterwards with the free-hand; let the two sides forming the right angle be respectively 6 and 8 inches long, the hypothenuse 10 inches; show that the square described on the hypothenuse is equal to the sum of the squares described on the other two sides, by dividing each of the three squares into small squares whose area is one inch each; then counting and comparing their number in the two smaller with the number comprised in the larger square. Of course this measuring by the comparison of equal units *proves* nothing logically. The *demonstration* of this proposition will come hereafter when the reasoning faculty is under discipline.

Draw with the proper compasses a large circle on the board; call attention to the fact that the curve bounding the circle you have made, is its circumference; point to the centre in which one leg of the compass is fixed, and show with the other that the circumference is everywhere at an equal distance from it; draw different straight lines from the centre to the circumference, name each a radius and show that the radii under inspection are equal to each other; from opposite points in the circumference, draw a straight line through the centre from left to right; name the line so drawn the diameter of the circle and point out that its diameter divides the circle into two equal parts—halves—or semicircles; connect opposite points in the circumference by a second straight line running through the centre perpendicular to the first; show that the two diameters (the perpendicular and the horizontal) together divide the circle into four parts—quarters or quadrants and make it plain by measurement that these quadrants are equal to each other. Finally divide a portion of the circumference into equal arcs by drawing different radii from the centre and illustrate, by the compasses, that equal arcs of the circumference subtend equal angles at the centre; then by drawing different radii so as to divide another portion of the circumference into unequal arcs, show in the same manner that they subtend unequal angles at the centre.

287. The Simultaneous Training of the Ear[1] **and the Tongue.**—Along with the drill that strengthens the early perceptions of touch and sight, should go corresponding exercises that give accuracy and readiness to the ear and the tongue. In the educating process the tongue is in a manner related to the ear as the hand is related to the eye. Both hand and tongue are instruments of expression; the one through lines and script, etc., the other through sounds and vocal language. The hand presents its products to the eye; the tongue presents its products to the ear. The former not only teaches the eye primarily the meaning of light and shade as indicating shape, but subsequently

---

[1] Chap. XVIII. 217.

increases its acuteness by reproducing its images in lines and colors. The latter incited to its first rudimentary utterances by the sounds which the ear perceives, becomes presently the principal means by which the ear is educated to greater keenness and rapidity of discernment. But an important distinction between the two is still to be noted. The hand being as we have seen, the organ of perception as well as sensation, may, in the absence of sight, be trained to a power of subtle discrimination that well nigh supplies its place. But the tongue,[1] being the organ of sensation only, will, in the absence of hearing, remain forever mute. It is evident then that the ear and the tongue must be trained in concert; the one by discriminating sounds attentively; the other by reproducing them distinctly.

288. The Sounds best Adapted to Training the Ear and the Tongue.—Now arises the question what sounds are most suitable for the early systematic training of the ear and the tongue? The answer is at hand. They should conform to rules corresponding to those given for the selection and arrangement of elementary forms. They must be distinct, moderate in volume, melodious if possible, mostly vocal and imitable by the juvenile tongue. In an arrangement for consecutive exercises, they should begin with the simplest vocal elements and rise gradually to their more complex combinations.

289. Musical Sounds. Among the sounds most available for primary lessons addressed to the ear and furnishing suitable vocal practice, the simple melodies hold a prominent place. They should consequently be employed·at frequent intervals in every series of exercises for the early training of the ear and the

---

[1] The term tongue here includes the entire vocal organs.

tongue. In the singing of simple airs, the ear grows susceptible to beauty of sound and the tongue becomes flexible and facile in their utterance. Another value they possess, must have its due weight in settling the comparative time given to the practice of elementary music. Being universally attractive to children it supplies the means of a happy relief to exercises that are less fascinating. Of course, for the purpose in hand, the elementary melodies should not be produced apart from words. The music of simple verse has a perpetual charm for the ear of the average girl or boy and the charm is redoubled when such verse is set to fitting music. There is, beyond question, a pressing demand for greater and better facilities in this line. We need urgently a more abundant supply of poetry and music that will meet fully the necessities of primary instruction. For no instrument is more effective in primary education than melodious sounds, produced in metrical language. The same heed should be given here to the rules so widely applicable in elementary training, namely, that the intervals of singing should be frequent and brief, that the voice should not be overstrained and that the class should be taught in concert by imitation only. Childhood is the age of imitation, which is accordingly a large element in elementary methods of instruction. A good voice in singing and a facile hand in drawing are well nigh indispensable to the primary teacher, as qualifications for his work.

290. The Elementary Articulate Sounds.—The early series of consecutive lessons in sounds inciting the ear and the tongue to systematic efforts, will comprise, in large measure, (1) The enunciation of each vowel and consonant sound; (2) Combining these vocal elements into the simplest syllables distinctly uttered; (3) Traversing, in this way, step by step, the entire succession of monosyllables, dissyllables, trisyllables, etc., until the pupil can spell phonetically and pronounce correctly any word which he distinctly hears. Of course abundant time must be given to these lessons in phonetics. Every sound must be learned in the first instance by imitating the teacher's utterance of it.

291. Training the Eye, Hand, Ear, and Tongue by Drawing and Naming the Capital Letters.— At the point when we closed the exercises in drawing the regular figures, we may fitly begin lessons in drawing the capital letters of the alphabet and learning incidently their names. For this purpose the capital letters may be happily divided into progressive models which begin with characters formed by the union of perpendicular and horizontal lines only. The next will be composed of exclusively the forms made by the joining of perpendicular, horizontal, and oblique lines at acute angles. The third group will include the capitals that contain the various curves.

The logical order of each exercise is substantially as follows: 1. The drawing of the capital letter. 2. The distinct and repeated utterance of its name. 3. The clear articulation of the sound or sounds it represents.

The early practice in writing is also the most effective means of teaching rudimentary reading. Character spelling should be learned wholly through the use of the pen or pencil. Here follow the lessons in which the pupil combines into the simplest syllables the letters he writes, then pronouncing their names, he finally spells phonetically and articulates distinctly, the spoken syllable for which they stand. In this way *sight* spelling and *sound* spelling are effectively united in the same exercises and made helpful to each other. By this method alone can our defective alphabet wherein twenty-four characters represent forty-two elementary sounds be thoroughly and practically learned by beginners. Moreover, besides the eye-training that the learner gains thereby and with which he finally conquers the entire orthography of the language, his tongue soon acquires habits of distinct utterance in reading and speaking that continue through life.

The daily practice in writing and spelling both by sight and by sound, may be profitably protracted from words of one to words of many syllables, until the pupil perceives the form and utters the sound, without hesitation. Thereafter he may continue the practice, with pen or pencil of character-spelling separately, up to the point where his eyes catches it at a glance, and his memory fixes indelibly, the precise literal form of every word he sees. Under such training carefully conducted in the beginning, every child who is not an idiot, will become ultimately a correct speller.[1]

292. The Training in Concrete Numbers Preparatory to the Study of Arithmetic.—As expressed in our last chapter, the earliest abstract ideas gained by the child are probably elementary notions of number and form. Of these two abstract qualities of things, number is the more simple and the lessons in concrete[2] objects representing number, will therefore stand first in the series of mathematical studies. They may indeed precede the drawing of regular forms as described above, and be given along with the color lessons. Exercises in the concrete operations of number will greatly assist the natural movement of primary abstraction and lend clearness and distinctness to the concepts of abstract numbers. Since the simplest and

---

[1] Find detailed directions for teaching reading and spelling through writing, in Parker's "Talks on Teaching," or Welch's "Talks on Psychology."

[2] Chap. XVI. 211.

most facile classifications are those of *identities*, the objects used to represent concrete units should be similar in size, form, and colors. Tens and hundreds may be indicated by different colors. The ordinary numerical frame is the best instrument for the first lessons in concrete counting and reckoning, but the final ones may be taken on a large frame fastened to the wall suitable for exercises in concert and displaying much more amply the different units of the decimal system.

I will only add further to the complete directions given in this branch of instruction by writers on primary teaching, that each step in the progressive operations should be frequently repeated, that the concrete course in numbers should be carried as far as the nature of the case will admit, and that, at this stage, precision and expertness are the object sought rather than the reasons of things.

293. Lessons on Irregular Forms in Botany and Zoology in Connection with Drawing.—The pencil is certainly a most effective help to the juvenile efforts that give to the hand and the eye, their elementary training. It will be found equally helpful in the subsequent study of the sciences whose purpose is to classify the concrete things whether of the organic or inorganic world. Botany and zoology, the first especially, will supply numerous objects whose forms are simple and attractive and whose outlines are easily traced with the pencil. The leaf forms will compose the beginning series of irregular figures which may be inspected and named as to their parts and then sketched as to their outlines. Obviously the simple forms as linear, lanceolate, elliptical, oblong, oval, spatulate, cuneate, etc., will be foremost, coming in the order in which they are named. Then follow the leaves with varying edges undulate, serrate, dentate, ciliate, etc., and finally the exercises on leaf forms may close with the compound forms, commencing with those that display the fewest leaflets and closing with those whose parts are most numerous.

This series of lessons on the leaf forms will be made more effective by the use of models which have been collected by the pupils with the teacher's help. Similar collections may be preserved for the months when leaves are out of season. As the specimens under inspection increase in complexity, all the organs and parts should be noted and named, the terms employed carefully explained, and finally the forms outlined and finished with the pencil.

294. Lessons on Forms in Zoology.—The enthusiastic teacher who is permitted to shape the course of instruction he gives, may now proceed with similar lessons on a series of available forms which are furnished by the animal kingdom. The objects selected for this purpose may also be arranged on the basis of their suitableness as models for drawing. Probably in the animal series the various shells disposed in the order suggested above, will come earliest into play. Next perhaps the bird forms may be studied and drawn from suitable mounted specimens, which after serving the full purpose, may be succeeded by a few of the smaller quadrupeds whose figures are such as can be most easily analyzed and outlined.

It is obvious that all lessons on the objects presented to the senses, can be given only by the aid of suitable collections. Happily the gathering and preparing of such collections under proper guidance by the children themselves, will furnish further exercise to the hand and the eye.

The views on incipient intellectual training expressed

in this closing chapter, are not only founded on a rigid analysis of the juvenile faculties, but have been verified by fifteen years of experience in the State Normal School of Michigan.

QUESTIONS ON CHAPTER XXI.

At what time should the formal education of the child begin? What is the character of the first voluntary efforts? What is the principal danger at the incipient stage of formal instruction? What should be the character of the instruction given at this period? Varying ages of children when systematic training may begin. What is the proper office of the play-school, or kindergarten? What object should be chosen for the first formal training of the hand? The intellectual senses under normal conditions are trained simultaneously. What are the objects that train the sense of sight exclusively? What are the first colors that can be used in training the eye? Give the method of procedure in scrutinizing these colors. Name and distinguish the primary, secondary, and the tertiary colors? Give the three steps in the process by which each color is exhaustively studied. How may the tints of the different colors be presented and taught? Also the shades? Give the method by which the tinges of the three classes of colors may be thoroughly learned. Why should the teacher refrain from explaining the philosophy of colors at this period? Give the formula for producing the compound colors. Along with the color lessons, what other exercises may be given for training the eye and the hand together? Give the five characteristics that should be found in the objects used for training the intellectual senses. Why is the pupil at this stage unable to gain an idea of these objects from definition? Give the four steps of the process of calling the eye, the tongue, and the hand into exercise on elementary forms. What simple equipment will be needed for giving primary lessons on form? What regular forms are the simplest and therefore most available for this purpose? Give the method by which the form of the simplest triangle may be taught through drawing. Also the method in drawing and naming the parts of a right-angled triangle. What other plane figures may be used for the same purpose? Method in drawing the square and parallelograms. Importance of scrupulous attention to each step. Give the steps which the pupil takes in drawing the three-square described on the sides of a right-angled triangle. The steps in drawing, describing and dividing the circle. Arranging and using the capital letters for drawing lessons. What are the reasons for the simultaneous training of the ear and the tongue? What is the character of the sounds best adapted to the training of the ear and the tongue? In what manner may musical sounds be made available for this purpose? Through what means should children be taught the simplest and earliest lessons in music? By what method may the articulate sounds be used in training the tongue and the ear of the child? When should the first lessons in writing be given? How may writing be made the instrument by which reading and spelling are taught? Describe the training in concrete numbers which is preparatory to the study of arithmetic. Describe the drawing lessons by which the elementary forms in botany may be taught. Also the lessons on the simplest forms in zoology.

INDEX.

Abstract Concept, The, 106
  Is what? 9
  How Evolved, 233
Abstract Ideas: Their Origin, 105
Abstraction: Its Nature 104
  Is what? 8
  How Trained, 206
  Appears Early, 105
Acquisition: Helped by Feeling, 69
                by Novelty, 69
                by Beauty, 68
Algebra as a Study, 244
Analysis: Is what? 8
  of Object, under Senses, 101
  Scientific and Ordinary, 99
  How Trained, 205
  Defined, 98
Appetites: what? 17
Arithmetic as a Study, 242
Artificial Language Addresses what? 216
Association: of Whole and Part, 77
  of Time, 74
  by Contrast, 81
  of Cause and Effect, 79
  of Sign and Thing, 78
  by Resemblance, 80
  by Place, 75
Attention: what? 5
  Importance of, 58
  And Observation, 58
  Related to Reflection, 58

Botany as a Study, 239

Cause: How Suggested, 172
Child Learns to Read, How, 225
Class Concept is what? 11, 90
Class Judgment: Its Employment, 134
Class Names Designated, 121
Classification: and Mental Power, 119
  Arranges Concepts, 119
  Is Arranging Subjects, 125
  Based on Form, 126
  and Language, 130
  Its Mental Process, 121
  Scientific: what it is, 124

Classification: in Children, 122
  Needs Analysis, 121
  Perfect, is what? 124
  Exemplified, 124
  of Concrete Objects: Example, 129
  Is what? 10
  Powers: How Educated, 209
  Its Preceding Acts, 119
  Desultory, is what? 122
Classifying Faculty, The, Needs Training, 186
Colors: Use of, 262
Composition as a Study, 248
Comparison, the Basis of Thinking, 135
Comprehension and Extension, Definition of, 130
Concept: Is what? 7
  of Substance, 170
  of Time: what is it? 167
  of Cause: what is it? 168
  of Nuomenon, 177
  of Phenomenon, 1777
  of Space is what? 165
  Depends on Percept, 100
  Objects Sometimes too Vivid, 94
Conception: Is what? 7
  Faculty of, 185
  Defined, 90
  How Trained, 205
Concepts: of Teacher must be Distinct, 95
  Arranged in Classes, 120
  Modified by Imagination, 91
  of Touch and Hearing, 93
  of Sight, 91
  of Hearing, 93
Conscience Defined, 22
Consciousness: Is what? 29
  and the Mental Act, 31
  a Present Knowledge, 30
  and the Mind, 31

Deductive Reasoning: Two Kinds, 148
Definition, A: what it is, 120
  Defined, 128

Descriptive Power Consists in what?
   96
Desire: what it is, 3
   Becomes Motive for Action, 2
Desires, The, Discussed, 24

EDUCATION: Defined, 184
   Principles of, 195
   Early Principles Consist in what?
      122
   Moral, 184
   Intellectual, 184
   Physical, 184
   Requires Useful Knowledge, 193
   When it Begins, 258
   Demands Reiteration, 192
   Its Effect on Analysis, 97
   Problems in, to be Solved, 191
   and Environment, 189
   Demands Strenuous Effort, 191
   Reaches all Faculties, 196
Emotions: what? 20
   of Beauty Explained, 21
   of Sublimity Explained, 21
Environment: Relation to Education, 189
   Must be Arranged, 189
English Grammar for Advanced Pupils only, 234
   As a Study, 249
Expression: in Natural Language, 214
   Defined, 214
   Effect of, 218
   in Artificial Language, 215
   and Education, 218
Eye and Ear Compared, 50

FACULTY: Is what? 4
   Defined, 155
   Its Object and Product, 155
Feeling: what it is, 2
   Its Effects on Conception, 94
   and Desire: How Related, 28
Feelings either Pleasant or Painful, 23

GENERAL Truths, Need of, in Deductive Reasoning, 150
Genius Defined, 115
Geometry as a Study, 245
Geology as a Study, 240
Geography, The Study of, 252
Grammar, a Subjective Science, 235

HAMILTON Quoted, 128
   Hand: Teaches the Eye, 49
Hand and Eye, Training of, 267
   Needs Training, 185, 199
Hearing Needs Training, 201

IDEAS Connected by Association, 84
   What ones can be Analyzed, 102
Ignorant, The, and Class Concepts, 123
Image Concepts: what? 10
Imagination: Is what? 9
   Defined, 108
   A Building Power, 108
   in the Fine Arts, 113
   in Poetry, 115
   and Subjective Poetry, 115
   and Objective Poetry, 115
   in the Arts, 112
   Effect on Character, 112
   Needs Training, 185
   A Wide Range of, 111
   Affected by Passion, 110
   Gets its Material from the Eye, 110
   Not an Originator, 109
   Often Distorts Facts, 110
   and Language, 117
   How Educated, 208
Internal Perception Defined, 56
Induction often Hasty, 153
Inductive Reasoning Defined, 152
Intuition: Constituent Power, 179
   Definition, 163
   Not yet Fathomed, 117
   Its Early Activity, 173
   in Savages, 179
   Susceptible of Culture, 179
   Must be Exercised, 179
Instruction must be in Accord with Mental Unfolding, 157
Instincts Defined, 18
Interest and Reiteration, 203

JUDGMENT: Is what? 11
   Defined, 134
   Is Thinking in Concepts, 144
   and Comparison, 137
   May be Trained, 187
   How Educated, 210
   in Extension and Comprehension, 142
   and the Proposition, 139
   Explained by Hamilton, 142
   Affirmative and Negative, 139

KNOWING: what it is, 2
   Knowledge: Consists in what? 157
   Related to Induction, 153
   Its Relation to Class Concepts, 123
   and Feeling, 15
   and Feeling are Joined, 37
   Needed in order to Feel, 16
   Tends to Fade, 70
   Its Renewal, 72

Index. 285

LANGUAGE Defined, 116
  And Classification, 221
  Its Value as a Gymnastic, 221
  As a Study, 246
Laws of Mind Growth, The, 236
Love of Knowledge, 22

MATHEMATICS Educate Reasoning Power, 212
Manual Training must be a First Step, 198
Memory: Is what? 6
  Value of, 63
  Its Development, 62
  Three Kinds, 64
  How Helped, 204
  How Trained, 120
  Trained by Vivid Concepts, 202
  Varieties of, 62
  Instances of, 63
  A Selecting Power, 71
  of Possible Knowledge, 70
  Demands Attention, 66
  Affected by Special Bent of Mind, 68
  Order of Action, 64
  Untrained, Yields Obscure Concepts, 188
Mental Operations: with Unfamiliar Objects, 161
  Rapidity of, 29
  Order of, 28
Mental Acts, Complex, 160
Mind: what it is, 2
  Compares Concepts, 136
  and Matter Contrasted, 14
  Manifestations: Three Kinds, 15
Mineralogy as a Study, 240
Moral Character: what is it? 28
Motives: True and False, 27
  Should be Directed by Higher Impulses, 27

NECESSARY Truths Defined, 180
  Noumenon, 174
    and Phenomenon, 174

OBJECTS of Intellectual Discipline, 229
Observation Defined, 59
Obstacles to Student's Progress, 235

PERCEPT is what? 6
  Perception is what? 5
Play Schools, Value of, 259
Premise and Conclusion, 151
Product of Faculty: what? 4
Proposition Defined, 140

REASONING: Is what? 11
  Defined, 145
  Powers: How Educated, 211
  Its Axioms, 145
  Deductive or Inductive, 145
  in Comprehension: what? 148
  And Comparison of Concepts, 146
Recollection Defined, 74
Reflection: Defined, 59
  Its Development, 60
Reiteration: Its Importance, 228
  Related to Spontaneity, 228
Religious Emotions, 23
Retention Defined, 70
  Is Spontaneous, 82
  and Association, 74
  and Knowledge Defined, 70
Retentive Power: Its Importance, 73
Reviewing and Retention, 73
Reviews, Value of, 227
Rhetoric as a Study, 250

SENSATION: Is what? 5
  Culture of, 44
  Gives Rise to Pleasure, 43
  and Perception Contrasted, 37
Sensations: Pleasant or Painful, 17
  Characterized by Pleasure, 44
  How Trained, 197
  and Perceptions Compared, 40
  and Perceptions in the Memory, 39
Selfish Feelings, like Appetites, 19
  Related to Education, 19
  What? 18
Sense Training is Memory Training, 203
Sense Concepts differ from Image Concepts, 109
Sense Concepts resemble Image Concepts, 110
Sense Perception: Three Kinds, 81
  and the Mind: How Related, 43
  and Sensation, 36
  Gives Rise to Names, 42
Senses: Six in Number, 39
  Order of Growth, 47
  Purpose of, 35, 42
  Intellectual, 48
  Organ of, 35
Sight Concepts: Their Superiority, 92
  Our Standards, 92
Sight Needs Training, 199
Social Feelings: what? 19
  Their Object, 19
  Their Relation to Civilization, 20
Spontaneities, The: How Produced, 223
  Directed by the Will, 224

Spontaneity: Is what? 3
  Begins all Knowledge, 82
Student's Progress, Obstacles to, 235
Studies: Course of, to be arranged:
    How? 190
  Suggested, 255
  to be Arranged by Teachers, 190
  Selected as Gymnastics, 238
Substance: How Suggested, 171
Syllogism Defined, 148

TASTE Defined, 115
  Thought, its Nature: How Learned, 137
Thought: Illustrated, 137
  and the Concept, 134
  Groups and Concepts, 135
  Moves in Judgments, 137
  Controlled by Will, 84
Time: How Suggested, 171
  Value of Knowledge for, 217
Teaching, Bad, Illustrated, 233
Teaching of Geography, 233
  of Botany, 279
  of Numbers, 278

Teaching of Language, 215
  of Zoology, 279
Term: Exact Meaning of, Necessary, 2
  Restricted to Object, 2
Thinking, Kinds of, Contrasted, 85
  in Judgment, 137
  Means Comparison, 135
Training: in Botany, 279
  of Ear and Tongue, 274
  of the Hand, 260
  in Numbers, 278
  in Natural Language, 215
  of the Sight, 261
  in Zoology, 279

VALUE of Language, 220
  of Knowledge for Training, 217

WILL, The: what it is, 3
  Is what? 24
  Means Effort Making, 3
  Preceded by Desire, 25
  in Relation to Feeling, 25
  Controls Recollection, 83
Willing means what? 3

THE SCHOOL JOURNAL

is published weekly at $2.50 a year and is in its 23rd year. It is the oldest, best known and widest circulated educational weekly in the U. S. THE JOURNAL is filled with *ideas* that will surely advance the teachers' conception of education. The best brain work on the work of professional teaching is found in it —not theoretical essays, nor pieces scissored out of other journals—THE SCHOOL JOURNAL has its own special writers— the ablest in the world.

THE PRIMARY SUPPLEMENT

of THE JOURNAL, is published in separate form monthly from September to June at $1.00. It is the ideal paper for primary teachers, being devoted almost exclusively to original primary methods and devices.

THE TEACHERS' INSTITUTE

is published monthly, at $1.25 a year; 12 large 44 page papers constitute a year—most other educational monthlies publish 10, some 9. It is edited in the same spirit and from the same standpoint as the JOURNAL, and has ever since it was started in 1878 been the *most popular educational monthly published*, circulating in every state. Every line is to the point. It is finely printed and crowded with illustrations made for it. Every study taught by the average teacher is covered in each issue.

EDUCATIONAL FOUNDATIONS.

This is *not* a paper: it is a series of small monthly volumes that bear on Professional Teaching. It is useful for those who want to study the foundations of education; for Normal Schools, Training Classes, Teachers' Institutes and individual teachers. If you desire to teach professionally you will want it. Handsome paper covers, 96 pp. each month. During 1892-93 Herbert Spencer's famous book on "Education" will be printed in it 32 pp. at a time. This alone is worth at least $1.00.

OUR TIMES.

Was started two years ago to give a resume of the important news of the month—not the murders, the scandals, etc., but *the news* that bears upon the progress of the world and specially written for the schoolroom. In Sept 1892 it was doubled in size, the 8 extra pages giving many fresh dialogues, recitations and declamations, and exercises for special days. This material alone during the year would cost at least 50 cents in the cheapest book form. Club rates, 40 cents.

*** *Select the paper suited to your needs and send for a free sample. Samples of all the papers 25 cents.*

E. L. KELLOGG & CO., New York and Chicago.

SEND ALL ORDERS TO
6   *E. L. KELLOGG & CO., NEW YORK & CHICAGO.*

Allen's Mind Studies for Young Teach-

ERS. By JEROME ALLEN, Ph.D., Associate Editor of the SCHOOL JOURNAL, Prof. of Pedagogy, Univ. of City of N. Y. 16mo, large, clear type, 128 pp. Cloth, 50 cents; *to teachers*, 40 cents; by mail, 5 cents extra.

JEROME ALLEN, Ph.D. Associate Editor of the *Journal* and *Institute*.

There are many teachers who know little about psychology, and who desire to be better informed concerning its principles, especially its relation to the work of teaching. For the aid of such, this book has been prepared. But it is not a psychology—only an introduction to it, aiming to give some fundamental principles, together with something concerning the philosophy of education. Its method is subjective rather than objective, leading the student to watch mental processes, and draw his own conclusions. It is written in language easy to be comprehended, and has many practical illustrations. It will aid the teacher in his daily work in dealing with mental facts and states.

To most teachers psychology seems to be dry. This book shows how it may become the most interesting of all studies. It also shows how to begin the knowledge of self. "We cannot know in others what we do not first know in ourselves." This is the key-note of this book. Students of elementary psychology will appreciate this feature of "Mind Studies."

ITS CONTENTS.

CHAP.
- I. How to Study Mind.
- II. Some Facts in Mind Growth.
- III. Development.
- IV. Mind Incentives.
- V. A few Fundamental Principles Settled.
- VI. Temperaments.
- VII. Training of the Senses.
- VIII. Attention.
- IX. Perception.
- X. Abstraction.
- XI. Faculties used in Abstract Thinking.

CHAP.
- XII. From the Subjective to the Conceptive.
- XIII. The Will.
- XIV. Diseases of the Will.
- XV. Kinds of Memory.
- XVI. The Sensibilities.
- XVII. Relation of the Sensibilities to the Will.
- XVIII. Training of the Sensibilities.
- XIX. Relation of the Sensibilities to Morality.
- XX. The Imagination.
- XXI. Imagination in its Maturity.
- XXII. Education of the Moral Sense.

SEND ALL ORDERS TO
E. L. KELLOGG & CO., NEW YORK & CHICAGO. 9

Browning's Educational Theories.

By OSCAR BROWNING, M.A., of King's College, Cambridge, Eng. No. 8 of *Reading Circle Library Series.* Cloth, 16mo, 237 pp. Price, 50 cents; *to teachers,* 40 cents; by mail, 5 cents extra.

This work has been before the public some time, and for a general sketch of the History of Education it has no superior. Our edition contains several new features, making it specially valuable as a text-book for Normal Schools, Teachers' Classes, Reading Circles, Teachers' Institutes, etc., as well as the student of education. These new features are: (1) Side-heads giving the subject of each paragraph; (2) each chapter is followed by an analysis; (3) a very full *new* index; (4) also an appendix on "Froebel," and the "American Common School."

OUTLINE OF CONTENTS.

I. Education among the Greeks—Music and Gymnastic Theories of Plato and Aristotle; II. Roman Education—Oratory; III. Humanistic Education; IV. The Realists—Ratich and Comenius; V. The Naturalists—Rabelais and Montaigne; VI. English Humorists and Realists—Roger Ascham and John Milton; VII. Locke; VIII. Jesuits and Jansenists; IX. Rousseau; X. Pestalozzi; XI. Kant, Fichte, and Herbart; XII. The English Public School; XIII. Froebel; XIV. The American Common School.

PRESS NOTICES.

Ed. Courant.—"This edition surpasses others in its adaptability to general use."

Col. School Journal.—"Can be used as a text-book in the History of Education."

Pa. Ed. News.—"A volume that can be used as a text-book on the History of Education."

School Education, Minn.—"Beginning with the Greeks, the author presents a brief but clear outline of the leading educational theories down to the present time."

Ed. Review, Can.—"A book like this introducing the teacher to the great minds that have worked in the same field, cannot but be a powerful stimulus to him in his work."

Calkins' Ear and Voice Training by

MEANS OF ELEMENTARY SOUNDS OF LANGUAGE. By N. A. CALKINS, Assistant Superintendent N. Y. City Schools; author of "Primary Object Lessons," "Manual of Object Teaching," "Phonic Charts," etc. Cloth. 16mo, about 100 pp. Price, 50 cents; *to teachers*, 40 cents; by mail, 5 cents extra.

An idea of the character of this work may be had by the following extracts from its *Preface:*

"The common existence of abnormal sense perception among school children is a serious obstacle in teaching. This condition is most obvious in the defective perceptions of sounds and forms. It may be seen in the faulty articulations in speaking and reading; in the inability to distinguish musical sounds readily; also in the common mistakes made in hearing what is said. . . .

"Careful observation and long experience lead to the conclusion that the most common defects in sound perceptions exist because of lack of proper training in childhood to develop this power of the mind into activity through the sense of hearing. It becomes, therefore, a matter of great importance in education, that in the training of children due attention shall be given to the development of ready and accurate perceptions of sounds.

"How to give this training so as to secure the desired results is a subject that deserves the careful attention of parents and teachers. Much depends upon the manner of presenting the sounds of our language to pupils, whether or not the results shall be the development in sound-perceptions that will *train the ear and voice* to habits of distinctness and accuracy in speaking and reading.

SUPT. N. A. CALKINS.

"The methods of teaching given in this book are the results of an extended experience under such varied conditions as may be found with pupils representing all nationalities, both of native and foreign born children. The plans described will enable teachers to lead their pupils to acquire ready and distinct perceptions through sense training, and cause them to know the sounds of our language in a manner that will give practical aid in learning both the spoken and the written language. The simplicity and usefulness of the lessons need only to be known to be appreciated and used."

SEND ALL ORDERS TO
E. L. KELLOGG & CO., NEW YORK & CHICAGO. 45

Teachers' Manuals Series.

Each is printed in large, clear type, on good paper. Paper cover, price 15 cents; *to teachers*, 12 cents; by mail, 1 cent extra.

J. G. FITCH, Inspector of the Training Colleges of England.

There is a need of small volumes—"Educational tracts," that teachers can carry easily and study as they have opportunity. The following numbers have been already published.

It should be noted that while our editions of such of these little books that are not written specially for this series are as low in price as any other, the side-heads, topics, and analyses inserted by the editor, as well as the excellent paper and printing, make them far superior in every way to any other edition.

We would suggest that city superintendents or conductors of institutes supply each of their teachers with copies of these little books. Special rates for quantities.

No. 1. *Fitch's Art of Questioning.*
By J. G. FITCH, M.A., author of "Lectures on Teaching." 38 pp.
Already widely known as the most useful and practical essay on this most important part of the teachers' lesson-hearing.

No. 2. *Fitch's Art of Securing Attention.*
By J. G. FITCH, M. A. 39 pp.
Of no less value than the author's "Art of Questioning."

No. 3. *Sidgwick's On Stimulus in School.*
By ARTHUR SIDGWICK, M.A. 43 pp.
"How can that dull, lazy scholar be pressed on to work up his lessons with a will?" This bright essay will tell how it can be done.

No. 4. *Yonge's Practical Work in School.*
By CHARLOTTE M. YONGE, author of "Heir of Redclyffe," 35 pp.
All who have read Miss Yonge's books will be glad to read of her views on School Work.

No. 5. *Fitch's Improvement in the Art of Teaching.*
By J. G. FITCH, M.A. 25 pp.
This thoughtful, earnest essay will bring courage and help to many a teacher who is struggling to do better work. It includes a course of study for Teachers' Training Classes.

No. 6. Gladstone's Object Teaching.
By J. H. GLADSTONE, of the London (Eng.) School Board. 25 pp.
A short manual full of practical suggestions on Object Teaching.

No. 7. Huntington's Unconscious Tuition.
Bishop Huntington has placed all teachers under profound obligations to him by writing this work. The earnest teacher has felt its earnest spirit, due to its interesting discussion of the foundation principles of education. It is wonderfully suggestive.

No. 8. Hughes' How to Keep Order.
By JAMES L. HUGHES, author of "Mistakes in Teaching."
Mr. Hughes is one of the few men who know what to say to help a young teacher. Thousands are to-day asking, "How shall we keep order?" Thousands are saying, "I can teach well enough, but I cannot keep order." To such we recommend this little book.

No. 9. Quick's How to Train the Memory.
By Rev. R. H. QUICK, author of "Educational Reformers."
This book comes from school-room experience, and is not a matter of theory. Much attention has been lately paid to increasing the power of memory. The teacher must make it part of his business to store the memory, hence he must know how to do it properly and according to the laws of the mind.

No. 10. Hoffman's Kindergarten Gifts.
By HEINRICH HOFFMAN, a pupil of Froebel.
The author sets forth very clearly the best methods of using them for training the child's senses and power of observation.

No. 11. Butler's Argument for Manual Training.
By NICHOLAS MURRAY BUTLER, Pres. of N. Y. College for Training of Teachers.
A clear statement of the foundation principles of Industrial Education.

No. 12. Groff's School Hygiene.
By Pres. G. G. GROFF, of Bucknell University, Pa.

No. 13. McMurry's How to Conduct the Recitation.
By CHAS. MCMURRY, Prof. in State Normal School, Winona, Minn.
In 34 pp. is explained the ideas of the Hubart school of educators as regards class teaching. These are now acknowledged to be the scientific method. Grube's plan for teaching primary arithmetic is in the same line.

No. 14. Carter's Artificial Production of Stupidity
IN SCHOOLS. By R. BRUDENELL CARTER, F. R. S.
This celebrated paper has been so often referred to that we reprint it in neat form, with side-headings. 49 pp.

No. 15. Kellogg's Pestalozzi :
HIS EDUCATIONAL WORK AND PRINCIPLES. By AMOS M. KELLOGG, editor of the *School Journal.* 29 pp.
A clear idea is given in this book of what this great reformer and discoverer in education thought and did. His foundation principles are made specially prominent.

No. 16. Lang's Basedow.
32 pp. Same price as above.

No 17. Lang's Comenius.
By OSSIAN H. LANG. 32 pp. Same price as above.

SEND ALL ORDERS TO
E. L. KELLOGG & CO., NEW YORK & CHICAGO. 11

Currie's Early Education.

"The Principles and Practice of Early and Infant School Education." By JAMES CURRIE, A. M., Prin. Church of Scotland Training College, Edinburgh. Author of "Common School Education," etc. With an introduction by Clarence E. Meleney, A. M., Supt. Schools, Paterson, N. J. Bound in blue cloth, gold, 16mo, 290 pp. Price, $1.25; *to teachers,* $1.00; by mail, 8 cents extra.

WHY THIS BOOK IS VALUABLE.

1. Pestalozzi gave New England its educational supremacy. The Pestalozzian wave struck this country more than forty years ago, and produced a mighty shock. It set New England to thinking. Horace Mann became eloquent to help on the change, and went up and down Massachusetts, urging in earnest tones the change proposed by the Swiss educator. What gave New England its educational supremacy was its reception of Pestalozzi's doctrines. Page, Philbrick, Barnard were all his disciples.

2. It is the work of one of the best expounders of Pestalozzi.

Forty years ago there was an upheaval in education. Pestalozzi's words were acting like yeast upon educators; thousands had been to visit his schools at Yverdun, and on their return to their own lands had reported the wonderful scenes they had witnessed. Rev. James Currie comprehended the movement, and sought to introduce it. Grasping the ideas of this great teacher, he spread them in Scotland; but that country was not elastic and receptive. Still, Mr. Currie's presentation of them wrought a great change, and he is to be reckoned as the most powerful exponent of the new ideas in Scotland. Hence this book, which contains them, must be considered as a treasure by the educator.

3. This volume is really a Manual of Principles of Teaching.

It exhibits enough of the principles to make the teacher intelligent in her practice. Most manuals give details, but no foundation principles. The first part lays a psychological basis—the only one there is for the teacher; and this is done in a simple and concise way. He declares emphatically that teaching cannot be learned empirically. That is, that one cannot watch a teacher and see *how* he does it, and then, imitating, claim to be a teacher. The principles must be learned.

4. It is a Manual of Practice in Teaching.

SEND ALL ORDERS TO
E. L. KELLOGG & CO., NEW YORK & CHICAGO.

Dewey's How to Teach Manners in the

SCHOOL-ROOM. By Mrs. JULIA M. DEWEY, Principal of the Normal School at Lowell, Mass., formerly Supt. of Schools at Hoosick Falls, N. Y. Cloth, 16mo, 104 pp. Price, 50 cents; *to teachers*, 40 cents; by mail, 5 cents extra.

Many teachers consider the manners of a pupil of little importance so long as he is industrious. But the boys and girls are to be fathers and mothers; some of the boys will stand in places of importance as professional men, and they will carry the mark of ill-breeding all their lives. Manners can be taught in the school-room: they render the school-room more attractive; they banish tendencies to misbehavior. In this volume Mrs. Dewey has shown how manners can be taught. The method is to present some fact of deportment, and then lead the children to discuss its bearings; thus they learn why good manners are to be learned and practised. The printing and binding are exceedingly neat and attractive."

OUTLINE OF CONTENTS.

Introduction.
General Directions.
Special Directions to Teachers.

LESSONS ON MANNERS FOR YOUNGEST PUPILS.
Lessons on Manners — Second Two Years.
Manners in School—First Two Years.
    "   "   Second   "
Manners at Home—First   "
    "   "   Second   "
Manners in Public—First   "
    "   "   Second   "

Table Manners—First Two Years.
    "   "   Second   "
LESSONS ON MANNERS FOR ADVANCED PUPILS.
Manners in School.
Personal Habits.
Manners in Public.
Table Manners.
Manners in Society.
Miscellaneous Items.
Practical Training in Manners.
Suggestive Stories, Fables, Anecdotes, and Poems.
Memory Gems.

Central School Journal.—"It furnishes illustrative lessons."
Texas School Journal.—"They (the pupils) will carry the mark of ill-breeding all their lives (unless taught otherwise)."
Pacific Ed. Journal.—"Principles are enforced by anecdote and conversation."
Teacher's Exponent.—"We believe such a book will be very welcome."
National Educator.—"Common-sense suggestions."
Ohio Ed. Monthly.—"Teachers would do well to get it."
Nebraska Teacher.—"Many teachers consider manners of little importance, but some of the boys will stand in places of importance."
School Educator.—"The spirit of the author is commendable."
School Herald.—"These lessons are full of suggestions."
Va. School Journal.—"Lessons furnished in a delightful style."
Miss. Teacher.—"The best presentation we have seen."
Ed. Courant.—"It is simple, straightforward, and plain."
Iowa Normal Monthly.—"Practical and well-arranged lessons on manners."
Progressive Educator.—"Will prove to be most helpful to the teacher who desires her pupils to be well-mannered."

14 E. L. KELLOGG & CO., NEW YORK & CHICAGO.

Fitch's Lectures on Teaching.

Lectures on Teaching. By J. G. FITCH, M.A., one of Her Majesty's Inspectors of Schools. England. Cloth, 16mo, 395 pp. Price, $1.25; *to teachers*, $1.00; by mail, postpaid.

Mr. Fitch takes as his topic the application of principles to the art of teaching in schools. Here are no vague and general propositions, but on every page we find the problems of the school-room discussed with definiteness of mental grip. No one who has read a single lecture by this eminent man but will desire to read another. The book is full of suggestions that lead to increased power.

1. These lectures are highly prized in England.
2. There is a valuable preface by Thos. Hunter, President of N. Y. City Normal College.
3. The volume has been at once adopted by several State Reading Circles.

EXTRACT FROM AMERICAN PREFACE.

"Teachers everywhere among English-speaking people have hailed Mr. Fitch's work as an invaluable aid for almost every kind of instruction and school organization. It combines the theoretical and the practical; it is based on psychology; it gives admirable advice on everything connected with teaching—from the furnishing of a school-room to the preparation of questions for examination. Its style is singularly clear, vigorous and harmonious."

Chicago Intelligence.—"All of its discussions are based on sound psychological principles and give admirable advice."

Virginia Educational Journal.—"He tells what he thinks so as to be helpful to all who are striving to improve."

Lynn Evening Item.—"He gives admirable advice."

Philadelphia Record.—"It is not easy to imagine a more useful volume."

Wilmington Every Evening.—"The teacher will find in it a wealth of help and suggestion."

Brooklyn Journal.—"His conception of the teacher is a worthy idea for all to bear in mind."

New England Journal of Education: "This is eminently the work of a man of wisdom and experience. He takes a broad and comprehensive view of the work of the teacher, and his suggestions on all topics are worthy of the most careful consideration."

Brooklyn Eagle: "An invaluable aid for almost every kind of instruction and school organization. It combines the theoretical and the practical; it is based on psychology; it gives admirable advice on everything connected with teaching, from the furnishing of a school-room to the preparation of questions for examination."

Toledo Blade: "It is safe to say, no teacher can lay claim to being well informed who has not read this admirable work. Its appreciation is shown by its adoption by several State Teachers' Reading Circles, as a work to be thoroughly read by its members."

SEND ALL ORDERS TO
E. L. KELLOGG & CO., NEW YORK & CHICAGO.

Kellogg's School Management:

"A Practical Guide for the Teacher in the School-Room." By AMOS M. KELLOGG, A.M. Sixth edition. Revised and enlarged. Cloth, 128 pp. Price, 75 cents ; *to teachers*, 60 cents ; by mail, 5 cents extra.

This book takes up the most difficult of all school work, viz.: the Government of a school, and is filled with original and practical ideas on the subject. It is invaluable to the teacher who desires to make his school a "well-governed" school.

1. It suggests methods of awakening an interest in the studies, and in school work. "The problem for the teacher," says Joseph Payne, "is to get the pupil to study." If he can do this he will be educated.

2. It suggests methods of making the school attractive. Ninety-nine hundredths of the teachers think young people should come to school anyhow ; the wise ones know that a pupil who wants to come to school will do something when he gets there, and so make the school attractive.

3. Above all it shows that the pupils will be self-governed when well governed. It shows how to develop the process of self-government.

4. It shows how regular attention and courteous behaviour may be secured.

5. It has an admirable preface by that remarkable man and teacher, Dr. Thomas Hunter, Pres. N. Y. City Normal College.

Home and School.—"Is just the book for every teacher who wishes to be a better teacher."

Educational Journal.—"It contains many valuable hints."

Boston Journal of Education.—"It is the most humane, instructive, original educational work we have read in many a day."

Wis. Journal of Education.—"Commends itself at once by the number of ingenious devices for securing order, industry, and interest."

Iowa Central School Journal.—"Teachers will find it a helpful and suggestive book."

Canada Educational Monthly.—"Valuable advice and useful suggestions."

Normal Teacher.—"The author believes the way to manage is to civilize, cultivate, and refine."

School Moderator.—"Contains a large amount of valuable reading; school government is admirably presented."

Progressive Teacher.—"Should occupy an honored place in every teacher's library."

Ed. Courant.—"It will help the teacher greatly.'

Va. Ed. Journal.—"The author draws from a large experience."

Hughes' Mistakes in Teaching.

BY JAMES J. HUGHES, Inspector of Schools, Toronto, Canada. Cloth, 16mo, 115 pp. Price, 50 cents; *to teachers*, 40 cents; by mail, 5 cents extra.

JAMES L. HUGHES, Inspector of Schools, Toronto, Canada.

Thousands of copies of the old edition have been sold. The new edition is worth double the old; the material has been increased, restated, and greatly improved. Two new and important Chapters have been added on "Mistakes in Aims," and "Mistakes in Moral Training." Mr. Hughes says in his preface: "In issuing a revised edition of this book, it seems fitting to acknowledge gratefully the hearty appreciation that has been accorded it by American teachers. Realizing as I do that its very large sale indicates that it has been of service to many of my fellow-teachers, I have recognized the duty of enlarging and revising it so as to make it still more helpful in preventing the common mistakes in teaching and training."

This is one of the six books recommended by the N. Y. State Department to teachers preparing for examination for State certificates.

CAUTION.

Our new AUTHORIZED COPYRIGHT EDITION, *entirely rewritten by the author, is the only one to buy. It is beautifully printed and handsomely bound. Get no other.*

CONTENTS OF OUR NEW EDITION.

CHAP. I. 7 Mistakes in Aim.
CHAP. II. 21 Mistakes in School Management.
CHAP. III. 24 Mistakes in Discipline.
CHAP. IV. 27 Mistakes in Method.
CHAP. V. 13 Mistakes in Moral Training.

☞ *Chaps. I. and V. are entirely new.*

Hughes' Securing and Retaining Atten-

TION. By JAMES L. HUGHES, Inspector Schools, Toronto, Canada, author of "Mistakes in Teaching." Cloth, 116 pp. Price, 50 cents; *to teachers*, 40 cents; by mail, 5 cents extra.

This valuable little book has already become widely known to American teachers. Our new edition has been almost *entirely re-written*, and several new important chapters added. It is the only AUTHORIZED COPYRIGHT EDITION. *Caution.*—Buy no other.

WHAT IT CONTAINS.

I. General Principles; II. Kinds of Attention; III. Characteristics of Good Attention; IV. Conditions of Attention; V. Essential Characteristics of the Teacher in Securing and Retaining Attention; VI. How to Control a Class; VII. Methods of Stimulating and Controlling a Desire for Knowledge; VIII. How to Gratify and Develop the Desire for Mental Activity; IX. Distracting Attention; X. Training the Power of Attention; XI. General Suggestions regarding Attention.

TESTIMONIALS.

S. P. Robbins, Pres. McGill Normal School, Montreal, Can., writes to Mr. Hughes:—"It is quite superfluous for me to say that your little books are admirable. I was yesterday authorized to put the 'Attention' on the list of books to be used in the Normal School next year. Crisp and attractive in style, and mighty by reason of its good, sound common-sense, it is a book that every teacher should know."

Popular Educator (Boston):—"Mr. Hughes has embodied the best thinking of his life in these pages."

Central School Journal (Ia.).—"Though published four or five years since, this book has steadily advanced in popularity."

Educational Courant (Ky.).—"It is intensely practical. There isn't a mystical, muddy expression in the book."

Educational Times (England).—"On an important subject, and admirably executed."

School Guardian (England).—"We unhesitatingly recommend it."

New England Journal of Education.—"The book is a guide and a manual of special value."

New York School Journal.—"Every teacher would derive benefit from reading this volume."

Chicago Educational Weekly.—"The teacher who aims at best success should study it."

Phil. Teacher.—"Many who have spent months in the school-room would be benefited by it."

Maryland School Journal.—"Always clear, never tedious."
Va. Ed. Journal.—"Excellent hints as to securing attention."
Ohio Educational Monthly.—"We advise readers to send for a copy."
Pacific Home and School Journal.—"An excellent little manual."
Prest. James H. Hoose, State Normal School, Cortland, N. Y., says:—"The book must prove of great benefit to the profession."
Supt. A. W. Edson, Jersey City, N. J., says:—"A good treatise has long been needed, and Mr. Hughes has supplied the want."

Payne's Lectures on the Science and

ART OF EDUCATION. *Reading Circle Edition.* By JOSEPH PAYNE, the first Professor of the Science and Art of Education in the College of Preceptors, London, England. With portrait. 16mo, 350 pp., English cloth, with gold back stamp. Price, $1.00 ; *to teachers*, 80 cents ; by mail, 7 cents extra. *Elegant new edition from new plates.*

JOSEPH PAYNE.

Teachers who are seeking to know the principles of education will find them clearly set forth in this volume. It must be remembered that principles are the basis upon which all methods of teaching must be founded. So valuable is this book that if a teacher were to decide to own but three works on education, this would be one of them. This edition contains all of Mr. Payne's writings that are in any other American abridged edition, and *is the only one with his portrait*. It is far superior to any other edition published.

WHY THIS EDITION IS THE BEST.

(1.) The *side-titles*. These give the contents of the page. (2.) The analysis of each lecture, with reference to the *educational* points in it. (3.) The general analysis pointing out the three great principles found at the beginning. (4.) The index, where, under such heads as Teaching, Education, The Child, the important utterances of Mr. Payne are set forth. (5.) Its handy shape, large type, fine paper, and press-work and tasteful binding. All of these features make this a most valuable book. To obtain all these features in one edition, it was found necessary to *get out this new edition*.

Ohio Educational Monthly.—"It does not deal with shadowy theories; it is intensely practical."

Philadelphia Educational News.—"Ought to be in library of every progressive teacher."

Educational Courant.—"To know how to teach, more is needed than a knowledge of the branches taught. This is especially valuable."

Pennsylvania Journal of Education.—"Will be of practical value to Normal Schools and Institutes"

Parker's Talks on Teaching.

Notes of "Talks on Teaching" given by COL. FRANCIS W. PARKER (formerly Superintendent of schools of Quincy, Mass.), before the Martha's Vineyard Institute, Summer of 1882. Reported by LELIA E. PATRIDGE. Square 16mo, 5x6 1-2 inches, 192 pp., *laid* paper, English cloth. Price $1.25; *to teachers*, $1.00; by mail, 9 cents extra.

The methods of teaching employed in the schools of Quincy, Mass., were seen to be the methods of nature. As they were copied and explained, they awoke a great desire on the part of those who could not visit the schools to know the underlying principles. In other words, Colonel Parker was asked to explain *why* he had his teachers teach thus. In the summer of 1882, in response to requests, Colonel Parker gave a course of lectures before the Martha's Vineyard Institute, and these were reported by Miss Patridge, and published in this book.

The book became famous; more copies were sold of it in the same time than of any other educational book whatever. The daily papers, which usually pass by such books with a mere mention, devoted columns to reviews of it.

The following points will show why the teacher will want this book.

1. It explains the "New Methods." There is a wide gulf between the new and the old education. Even school boards understand this.

2. It gives the underlying principles of education. For it must be remembered that Col. Parker is not expounding *his* methods, but the methods of nature.

3. It gives the ideas of man who is evidently an "educational genius," a man born to understand and expound education. We have few such; they are worth everything to the human race.

4. It gives a biography of Col. Parker. This will help the teacher of education to comprehend the man and his motives.

5. It has been adopted by nearly every State Reading Circle

Patridge's "Quincy Methods."

The "Quincy Methods," illustrated; Pen photographs from the Quincy schools. By LELIA E. PATRIDGE. Illustrated with a number of engravings, and two colored plates. Blue cloth, gilt, 12mo, 686 pp. Price, $1.75; *to teachers*, $1.40; by mail, 13 cents extra.

When the schools of Quincy, Mass., became so famous under the superintendence of Col. Francis W. Parker, thousands of teachers visited them. Quincy became a sort of "educational Mecca," to the disgust of the routinists, whose schools were passed by. Those who went to study the methods pursued there were called on to tell what they had seen. Miss Patridge was one of those who visited the schools of Quincy; in the Pennsylvania Institutes (many of which she conducted), she found the teachers were never tired of being told how things were done in Quincy. She revisited the schools several times, and wrote down what she saw; then the book was made.

1. This book presents the actual practice in the schools of Quincy. It is composed of "pen photographs."
2. It gives abundant reasons for the great stir produced by the two words "Quincy Methods." There are reasons for the discussion that has been going on among the teachers of late years.
3. It gives an insight to principles underlying real education as distinguished from book learning.
4. It shows the teacher not only what to do, but gives the way in which to do it.
5. It impresses one with the *spirit* of the Quincy schools.
6. It shows the teacher how to create an *atmosphere* of happiness, of busy work, and of progress.
7. It shows the teacher how not to waste her time in worrying over disorder.
8. It tells how to treat pupils with courtesy, and get courtesy back again.
9. It presents four years of work, considering Number, Color, Direction, Dimension, Botany, Minerals, Form, Language, Writing, Pictures, Modelling, Drawing, Singing, Geography, Zoology, etc., etc.
10. There are 686 pages; a large book devoted to the realities of school life, in realistic descriptive language. It is plain, real, not abstruse and uninteresting.
11. It gives an insight into real education, the education urged by Pestalozzi, Froebel, Mann, Page, Parker, etc.

Perez's First Three Years of Childhood.

AN EXHAUSTIVE STUDY OF THE PSYCHOLOGY OF CHILDREN. By BERNARD PEREZ. Edited and translated by ALICE M. CHRISTIE, translator of "Child and Child Nature," with an introduction by JAMES SULLY, M.A., author of "Outlines of Psychology," etc. 12mo, cloth, 324 pp. Price, $1.50; to teachers, $1.20; by mail, 10 cents extra.

This is a comprehensive treatise on the psychology of childhood, and is a practical study of the human mind, not full formed and equipped with knowledge, but as nearly as possible, *ab origine*—before habit, environment, and education have asserted their sway and made their permanent modifications. The writer looks into all the phases of child activity. He treats exhaustively, and in bright Gallic style, of sensations, instincts, sentiments, intellectual tendencies, the will, the faculties of æsthetic and moral senses of young children. He shows how ideas of truth and falsehood arise in little minds, how natural is imitation and how deep is credulity. He illustrates the development of imagination and the elaboration of new concepts through judgment, abstraction, reasoning, and other mental methods. It is a book that has been long wanted by all who are engaged in teaching, and especially by all who have to do with the education and training of children.

This edition has a new index of special value, and the book is carefully printed and elegantly and durably bound. Be sure to get our standard edition.

OUTLINE OF CONTENTS.

CHAP.
I. Faculties of Infant before Birth—First Impression of New-born Child.
II. Motor Activity at the Beginning of Life—at Six Months—at Fifteen Months.
III. Instinctive and Emotional Sensations—First Perceptions.
IV. General and Special Instincts.
V. The Sentiments.
VI. Intellectual Tendencies—Veracity—Imitation—Credulity.
VII. The Will.
VIII. Faculties of Intellectual Acquisition and Retention—Attention—Memory.

CHAP.
IX. Association of Psychical States—Association—Imagination.
X. Elaboration of Ideas—Judgment—Abstraction—Comparison—Generalization—Reasoning—Errors and Allusions—Errors and Allusions Owing to Moral Causes.
XI. Expression and Language.
XII. Æsthetic Senses—Musical Sense—Sense of Material Beauty—Constructive Instinct—Dramatic Instinct.
XIII. Personality—Reflection—Moral Sense.

Col. Francis W. Parker, Principal Cook County Normal and Training School, Chicago, says:—"I am glad to see that you have published Perez's wonderful work upon childhood. I shall do all I can to get everybody to read it. It is a grand work."

John Bascom, Pres. Univ. of Wisconsin, says:—"A work of marked interest."

G. Stanley Hall, Professor of Psychology and Pedagogy, Johns Hopkins Univ., says:—"I esteem the work a very valuable one for primary and kindergarten teachers, and for all interested in the psychology of childhood."

And many other strong commendations.

SEND ALL ORDERS TO
E. L. KELLOGG & CO., NEW YORK AND CHICAGO. 37

Shaw's National Question Book.

"THE NATIONAL QUESTION BOOK." A graded course of study for those preparing to teach. By EDWARD R. SHAW, Principal of the High School, Yonkers, N. Y., author of "School Devices," etc. Bound in durable English buckram cloth, with beautiful side-stamp. 12mo, 400 pp. Price, $1.75; *net to teachers*, postpaid.

A new edition of this popular book is now ready, containing the following

NEW FEATURES:

READING. An entirely new chapter with answers.

ALCOHOL and its effects on the body. An entirely new chapter with answers.

THE PROFESSIONAL GRADE has been entirely rewritten and now contains answers to every question.

This work contains 6,500 Questions and Answers on 24 Different Branches of Study.

ITS DISTINGUISHING FEATURES.

1. It aims to make the teacher a BETTER TEACHER.

"How to Make Teaching a Profession" has challenged the attention of the wisest teacher. It is plain that to accomplish this the teacher must pass from the stage of a knowledge of the rudiments, to the stage of somewhat extensive acquirement. There are steps in this movement; if a teacher will take the first and see what the next is, he will probably go on to the next, and so on. One of the reasons why there has been no movement forward by those who have made this first step, is that there was nothing marked out as a second step.

2. This book will show the teacher how to go forward.

In the preface the course of study usually pursued in our best normal schools is given. This proposes four grades; third, second, first, and professional. Then, questions are given appropriate for each of these grades. Answers follow each section. A teacher will use the book somewhat as follows :—If he is in the third grade he will put the questions found in this book concerning numbers, geography, history, grammar, orthography, and theory and practice of teaching to himself and get out the answer. Having done this he will go on to the other grades in a similar manner. In this way he will know as to his fitness to pass an examination for

Shaw and Donnell's School Devices.

"School Devices." A book of ways and suggestions for teachers. By EDWARD R. SHAW and WEBB DONNELL, of the High School at Yonkers, N. Y. Illustrated. Dark-blue cloth binding, gold, 16mo, 224 pp. Price, $1.25 ; *to teachers*, $1.00 ; by mail, 9 cents extra.

☞ A BOOK OF "WAYS" FOR TEACHERS. ☜

Teaching is an art; there are "ways to do it." This book is made to point out "ways," and to help by suggestions.

1. It gives "ways" for teaching Language, Grammar, Reading, Spelling, Geography, etc. These are in many cases novel; they are designed to help attract the attention of the pupil.

2. The "ways" given are not the questionable "ways" so often seen practiced in school-rooms, but are in accord with the spirit of modern educational ideas.

3. This book will afford practical assistance to teachers who wish to keep their work from degenerating into mere routine. It gives them, in convenient form for constant use at the desk, a multitude of new ways in which to present old truths. The great enemy of the teacher is want of interest. Their methods do not attract attention. There is no teaching unless there is *attention*. The teacher is too apt to think there is but one "way" of teaching spelling; he thus falls into a rut. Now there are many "ways" of teaching spelling, and some "ways" are better than others. Variety must exist in the school-room; the authors of this volume deserve the thanks of the teachers for pointing out methods of obtaining variety without sacrificing the great end sought—scholarship. New "ways" induce greater effort, and renewal of activity.

4. The book gives the result of large actual experience in the school-room, and will meet the needs of thousands of teachers, by placing at their command that for which visits to other schools are made, institutes and associations attended, viz., new ideas and fresh and forceful ways of teaching. The devices given under Drawing and Physiology are of an eminently practical nature, and cannot fail to invest these subjects with new interest. The attempt has been made to present only devices of a practical character.

5. The book suggests "ways" to make teaching *effective* ; it is not simply a book of new "ways," but of "ways" that will produce good results.

Tate's Philosophy of Education.

The Philosophy of Education. By T. TATE. Revised and Annotated by E. E. SHEIB, Ph.D., Principal of the Louisiana State Normal School. Unique cloth binding, laid paper, 331 pp. Price, $1.50; *to teachers*, $1.20; by mail, 7 cents extra.

There are few books that deal with the Science of Education. This volume is the work of a man who said there were great principles at the bottom of the work of the despised schoolmaster. It has set many a teacher to thinking, and in its new form will set many more.

Our edition will be found far superior to any other in every respect. The annotations of Mr. Sheib are invaluable. The more important part of the book are emphasized by leading the type. The type is clear, the size convenient, and printing, paper, and binding are most excellent.

Mr. Philbrick so long superintendent of the Boston schools hold this work in high esteem.

Col. F. W. Parker strongly recommends it.

Jos. MacAlister, Supt. Public Schools, Philadelphia, says:—"It is one of the first books which a teacher deserves of understanding the scientific principles on which his work rests should study."

Graded Examination Questions.

For N. Y. State, from Sept., 1887, to Sept., 1889, *with answers complete*. First, Second, and Third Grades. Cloth, 12mo, 219 pp. Price, $1.00; *to teachers*, 80 cents; by mail, 8 cents extra.

This volume contains the Uniform Graded Examination Questions, issued to the School Commissioners of the State by the Dept. of Public Instruction, commencing Sept., 1887, and ending Aug. 13 and 14, 1889. The answers are also given. These questions have been adopted by all the school commissioners of the State; the test in each county thus becomes uniform. These questions are being used very largely in many other States, that pattern after New York, and will therefore be of far more than local interest. Indeed, teachers and school officers in all States are using these questions as a basis for their own examinations. Our edition is the best in arrangement, print, binding, and has an excellent contents and index.

This book may be used to the best advantage by the teacher who desires to advance in the profession, because the questions are carefully graded. After the lowest grade of questions have been successfully answered, the next higher grade is studies. In our edition the answers are entirely separate from the questions in the back of the book.

Welch's Talks on Psychology Applied to

TEACHING. By A. S. WELCH, LL.D., Ex-Pres. of the Iowa Agricultural College at Ames, Iowa. Cloth, 16mo, 136 pp. Price, 50 cents; to *teachers*, 40 cents; by mail, 5 cents extra.

This little book has been written for the purpose of helping the teacher in doing more effective work in the school-room. The instructors in our schools are familiar with the branches they teach, but deficient in knowledge of the mental powers whose development they seek to promote. But no proficiency that does not include the *study of mind*, can ever qualify for the work of teaching. The teacher must comprehend fully not only the *objects* studied by the learner, but the *efforts* put forth and in studying them, the *effect* of these efforts on the faculty exerted, their *results* in the form of accurate knowledge. It is urged by eminent educators everywhere that a knowledge of the branches to be taught, and a *knowledge of the mind* to be trained thereby, are equally essential to successful teaching.

WHAT IT CONTAINS.

PART I.—Chapter 1. Mind Growth and its Helps. Chapter 2.—The Feelings. Chapter 3.—The Will and the Spontaneities. Chapter 4.—Sensation. Chapter 5.—Sense Perception, Gathering Concepts. Chapter 6.—Memory and Conception. Chapter 7.—Analysis and Abstraction. Chapter 8.—Imagination and Classification.—Chapter 9.—Judgment and Reasoning, the Thinking Faculties.

PART II.—Helps to Mind Growth. Chapter 1.—Education and the Means of Attaining it. Chapter 2.—Training of the Senses. Chapter 3.—Reading, Writing, and Spelling. Chapter 4.—Composition, Elementary Grammar, Abstract Arithmetic, etc.

* This book, as will be seen from the contents, deals with the subject differently from Dr. Jerome Allen's "Mind Studies for Young Teachers," (same price) recently published by us.

FROM THOSE WHO HAVE SEEN IT.

Co. Insp. Dearness, London, Canada.—"Here find it the most lucid and practical introduction to mental science I have ever seen."

Florida School Journal.—"Is certainly the best adapted and most desirable for the mass of teachers."

Penn. School Journal.—"Earnest teachers will appreciate it."

Danville, Ind., Teacher and Examiner.—"We feel certain this book has a mission among the primary teachers."

Iowa Normal Monthly.—"The best for the average teacher."

Prof. H. H. Seeley, Iowa State Normal School.—"I feel that you have done a very excellent thing for the teachers. Am inclined to think we will use it in some of our classes."

Science, N. Y.—"Has been written from an educational point of view."

Education, Boston.—"Aims to help the teacher in the work of the school-room."

Progressive Teacher.—"There is no better work."

Ev-Gov. Dysart, Iowa.—"My first thought was, 'What a pity it could not be in the hands of every teacher in Iowa.'"

Woodhull's Simple Experiments for the

School-Room. By Prof. JOHN F. WOODHULL, Prof. of Natural Science in the College for the Training of Teachers, New York City, author of "Manual of Home-Made Apparatus." Cloth, 16mo. Price, 50 cents; *to teachers*, 40 cents; by mail, 5 cents extra.

This book contains a series of simple, easily-made experiments, to perform which will aid the comprehension of every-day phenomena. They are really the very lessons given by the author in the Primary and Grammar Departments of the Model School in the College for the Training of Teachers, New York City.

The apparatus needed for the experiments consists, for the most part, of such things as every teacher will find at hand in a schoolroom or kitchen. The experiments are so connected in logical order as to form a continuous exhibition of the phenomena of *combustion. This book is not a science catechism.* Its aim is to train the child's mind in habits of reasoning by experimental methods.

These experiments should be made in every school of our country, and thus bring in a scientific method of dealing with nature. The present method of cramming children's minds with isolated facts of which they can have no adequate comprehension is a ruinous and unprofitable one. This book points out the method employed by the *best teachers in the best schools.*

WHAT IT CONTAINS.

I. Experiments with Paper.
II. " " Wood.
III. " " a Candle.
IV. " " Kerosene.
V. Kindling Temperature.
VI. Air as an Agent in Combustion.
VII. Products of Complete "
VIII. Currents of Air, etc.—Ventila-
IX. Oxygen of the Air. [tion.
X. Chemical Changes.

In all there are 91 experiments described, illustrated by 35 engravings.

Jas. H. Canfield, Univ. of Kans., Lawrence, says:—"I desire to say most emphatically that the method pursued is the only true one in all school work. Its spirit is admirable. We need and must have far more of this instruction."

J. C. Packard, Univ. of Iowa, Iowa City, says:—"For many years shut up to the simplest forms of illustrative apparatus, I learned that the necessity was a blessing, since so much could be accomplished by home-made apparatus—inexpensive and effective."

Henry R. Russell, Woodbury, N. J., Supt. of the Friends School:—"Admirable little book. It is just the kind of book we need."

S. T. Dutton, Supt. Schools, New Haven, Ct.—"Contains just the kind of help teachers need in adapting natural science to common schools."

SEND ALL ORDERS TO
E. L. KELLOGG & CO., NEW YORK & CHICAGO. 51

Wilhelm's Student's Calendar.

Compiled by N. O. WILHELM. Bound in paper. 76 pp. Double Indexed. Price, 30 cents; *to teachers*, 24 cents; by mail, 3 cents extra.

This is a perpetual calendar and book of days. It consists of Short Biographies of Greatest Men, arranged according to Birthdays and Deathdays, covering every day of the year.

These can be used for opening exercises in schools, for memorial days, and for giving pupils some information about the great men of the world about whom everybody ought to know something. Just the thing for families where there are young people.

The condensed information in this little book would in other form cost you many dollars to own. Here are a few of the names of persons whose Biographies are found in the "Student's Calendar:"

John Adams,	Queen Elizabeth,	John Hancock,	Abraham Lincoln,
J. Q. Adams,	R. W. Emerson,	Hamilton,	Jenny Lind,
Joseph Addison,	Robert Emmet,	Hannibal,	Linnæus,
Alexand'r the Gre't,	Euripides,	W. H. Harrison,	Dr. Livingstone,
Michael Angelo,	Edw. Everett,	Nath. Hawthorne,	H. W. Longfellow,
Aristotle,	Faraday,	Hayden,	Lowell,
Ascham,	Farragut,	Mrs. Hemans,	Lubbock,
Audubon,	Fénelon,	T. A. Hendricks,	Martin Luther,
Francis Bacon,	M. Fillmore,	Patrick Henry,	Macaulay,
Geo. Bancroft,	Chas. J. Fox,	Sir Wm. Herschel,	Macready,
Venerable Bede,	Ben. Franklin,	O. W. Holmes,	Mohammed,
Von Bismarck,	Sir J. Franklin,	Thomas Hood,	Horace Mann,
Tycho Brahe,	Frederick the Great	Jos. Hooker,	Maria Theresa,
Lord Brougham,	J. C. Fremont,	Horace,	Marie Antoinette,
Mrs. Browning,	Frobisher,	Sam. Houston,	Mary, Qu'n of Scots,
W. C. Bryant,	Froebel,	Elias Howe,	J. Montgomery,
Edmund Burke,	Froude,	Victor Hugo,	Sir J. Moore,
Robert Burns,	Robert Fulton,	Humboldt,	Mozart,
Ben. F. Butler,	Galileo,	David Hume,	Napoleon L,
Lord Byron,	Vasco da Gama,	Wash. Irving,	Nelson,
Cæsar,	Gambetta,	Andrew Jackson,	Sir Isaac Newton,
John Calhoun,	Garfield,	Jacotot,	Daniel O'Connell,
Thos. Campbell,	Garibaldi,	Jos. Jacquard,	Charles O'Conor,
Thos. Carlyle,	D. Garrick,	James I.,	Thos. Paine,
Phœbe Cary,	Horatio Gates,	James II.,	Geo. Peabody,
Cervantes,	R. Gatling,	John Jay,	Wm. Penn,
Salmon P. Chase,	George III.,	Thos. Jefferson,	Peter the Great,
Thos. Chatterton,	Stephen Girard,	Francis Jeffrey,	Pizarro,
Rufus Choate,	Gladstone,	Dr. Ed. Jenner,	Plato,
Cicero,	Goethe,	Joan of Arc,	E. A. Poe,
Henry Clay,	Goldsmith,	Sam'l Johnson,	W. H. Prescott,
Cleopatra,	U. S. Grant,	John Paul Jones,	Pulaski,
Coleridge,	Henry Grattan,	Dr. Kane,	Queen Victoria,
Schuyler Colfax,	Asa Gray,	John Keats,	Richelieu,
Anthony Collins,	Horace Greeley,	John Kitto,	J. P. Richter,
Cornwallis,	Nath. Greene,	Henry Knox,	Ritter,

Lubbock's Best 100 Books.

By Sir JOHN LUBBOCK. 64 pages, paper. Price, 20 cents; *to teachers*, 16 cents; by mail, 2 cents extra.

Sir John Lubbock, in an address last year before the Workingmen's College of London, England, gave a list of what he deemed the Best 100 Books. He said, in giving his list, that if a few good guides would draw up similar lists, it would be most useful.

The *Pall Mall Gazette* published Sir John Lubbock's list, and invited eminent men in England to give their opinions concerning it. We have just reprinted them in neat pamphlet form. Gladstone, Stanley, Black, and many others are represented.

SEND ALL ORDERS TO
E. L. KELLOGG & CO., NEW YORK & CHICAGO.   29

Reception Day. 6 Nos.

A collection of fresh and original dialogues, recitations, declamations, and short pieces for practical use in Public and Private Schools. Bound in handsome, new paper cover, 160 pages each, printed on laid paper. Price 30 cents each; *to teachers,* 24 cents; by mail, 3 cents extra.

The exercises in these books bear upon education; have a relation to the school-room.

NEW COVER.

1. The dialogues, recitations, and declamations, gathered in this volume being fresh, short, easy to be comprehended and are well fitted for the average scholars of our schools.
2. They have mainly been used by teachers for actual school exercises.
3. They cover a different ground from the speeches of Demosthenes and Cicero—which are unfitted for boys of twelve to sixteen years of age.
4. They have some practical interest for those who use them.
5. There is not a vicious sentence uttered. In some dialogue books profanity is found, or disobedience to parents encouraged, or lying laughed at. Let teachers look out for this.
6. There is something for the youngest pupils.
7. "Memorial Day Exercises" for Bryant, Garfield, Lincoln, etc., will be found.
8. Several Tree Planting exercises are included.
9. The exercises have relation to the school-room and bear upon education.
10. An important point is the freshness of these pieces. Most of them were written expressly for this collection, and *can be found nowhere else.*

Boston Journal of Education.—"Is of practical value."
Detroit Free Press.—"Suitable for public and private schools."
Western Ed. Journal.—"A series of very good selections."

WHAT EACH NUMBER CONTAINS.

No. 1

Is a specially fine number. One dialogue in it, called "Work Conquers," for 11 girls and 6 boys, has been given hundreds of times, and is alone worth the price of the book. Then there are 21 other dialogues.
29 Recitations.
14 Declamations.
17 Pieces for the Primary Class.

No. 2 Contains

20 Recitations.
12 Declamations.
17 Dialogues.
24 Pieces for the Primary Class.
And for Class Exercise as follows:
The Bird's Party.
Indian Names.
Valedictory.
Washington's Birthday.
Garfield Memorial Day.
Grant   "   "
Whittier "   "
Sigourney "   "

No. 3 Contains

Fewer of the longer pieces and more of the shorter, as follows:
18 Declamations.
21 Recitations.
22 Dialogues.
24 Pieces for the Primary Class.
A Christmas Exercise.
Opening Piece, and
An Historical Celebration.

No. 4 Contains

Campbell Memorial Day.
Longfellow   "   "
Michael Angelo "   "
Shakespeare.   "   "
Washington   "   "
Christmas Exercise.
Arbor Day   "
New Planting   "
Thanksgiving   "
Value of Knowledge Exercise.
Also 8 other Dialogues.
21 Recitations.
23 Declamations.

No. 5 Contains

Browning Memorial Day.
Autumn Exercise.
Bryant Memorial Day.
New Planting Exercise.
Christmas Exercise.
A Concert Exercise.
24 Other Dialogues.
16 Declamations, and
36 Recitations.

No. 6 Contains

Spring; a flower exercise for very young pupils.
Emerson Memorial Day.
New Year's Day Exercise.
Holmes' Memorial Day.
Fourth of July Exercise.
Shakespeare Memorial Day.
Washington's Birthday Exercise.
Also 6 other Dialogues.
6 Declamations.
41 Recitations.
15 Recitations for the Primary Class.
And 4 Songs.

Our RECEPTION DAY Series is not sold largely by booksellers, who, if they do not keep it, try to have you buy something else similar, but not so good. Therefore send direct to the publishers, by mail, the price as above, in stamps or postal notes, and your order will be filled at once. Discount for quantities.

SPECIAL OFFER.

If ordered at one time, we will send postpaid the entire 6 Nos. for $1.40. Note the reduction.

SEND ALL ORDERS TO
E. L. KELLOGG & CO., NEW YORK & CHICAGO.

Reinhart's Outline History of Education.

With chronological Tables, Suggestions, and Test Questions. By J. A. REINHART, Ph. D. Teachers' Professional Library. 77 pp., limp cloth, 25 cents; *to teachers*, 20 cents; by mail 2 cents extra.

This is one of the little books intended to be studied in connection with THE TEACHERS' PROFESSION. The publishers, by means of these publications bring to the very doors of those teachers who lack the opportunity to attend a normal school a chance to improve in the art of teaching. "Outlines of History of Education" is what its name implies, a brief but comprehensive presentation of the main facts in educational progress. The chapters are: Introduction; Education among the Greeks; Education among the Romans; Education in the Middle Ages; the Dawn of the New Era; Education and the Reformation; Education in the Seventeenth Century; Education in the Eighteenth Century; Education in the Nineteenth Century. A thorough study of this book will be a good foundation for a more detailed study of the subject. The book is well printed from clear, large type, with topic heads and questions, and is durably bound in limp cloth.

Reinhart's Outline Principles of Education

By J. A. REINHART., Ph. D. Teachers' Professional Library. 68 pp., limp cloth, 25 cents.

To give an outline of a great subject, including nothing trivial and leaving out nothing important, is a great art. This difficult task has been successfully performed by the author of this small volume, who is an educator of long experience, and a thorough student of the science of education. The first two chapters give a general view of the subject, and the other chapters treat of the intuitive, imaginative, and logical stages of education, and the principles of moral education. This is one of the volumes intended to be studied in connection with the monthly paper, THE TEACHERS' PROFESSION. Type, printing, binding are neat and durable, and like the History by same author.

REINHART'S CIVICS IN EDUCATION,

is another little book of same price and number of pages. Ready Nov. 1891.

SEND ALL ORDERS TO
E. L. KELLOGG & CO., NEW YORK & CHICAGO.

Quick's Educational Reformers.

By Rev. ROBERT HERBERT QUICK, M. A., of Trinity College, Cambridge, England. Bound in plain, but elegant cloth binding. 16mo, about 350 pp. $1.00; *to teachers*, 80 cts.; by mail, 10 cts. extra.

New edition with topical headings, chronological table and other aids for systematic study in normal schools and reading-circles.

No book in the history of education has been so justly popular as this. Mr. Quick has the remarkable faculty of grasping the salient points of the work of the great educators, and restating their ideas in clear and vigorous language.

This book supplies information that is contained in no other single volume, touching the progress of education in its earliest stages after the revival of learning. It is the work of a practical teacher, who supplements his sketches of famous educationists with some well-considered observations, that deserve the attention of all who are interested in that subject. Beginning with Roger Ascham, it gives an account of the lives and schemes of most of the great thinkers and workers in the educational field, down to Herbert Spencer, with the addition of a valuable appendix of thoughts and suggestions on teaching. The list includes the names of Montaigne, Ratich, Milton, Comenius, Locke, Rousseau, Basedow Pestalozzi. and Jacotot. In the lives and thoughts of these eminent men is presented the whole philosophy of education, as developed in the progress of modern times.

This book has been adopted by nearly every state reading-circle in the country, and purchased by thousands of teachers, and is used in many normal schools.

Contents. 1. Schools of the Jesuits; 2. Ascham, Montaigne, Ratich, Milton: 3. Comenius; 4. Locke· 5. Rousseau's Emile; 6. Basedow and the Philanthropin; 7. Pestalozzi; 8. Jacotot; 9. Herbert Spencer; 10. Thoughts and Suggestions about Teaching Children; 11. Some Remarks about Moral and Religious Education; 12. Appendix.

OUR NEW EDITION.

Be sure to get E. L. Kellogg's edition. There are other editions in the market that are not only higher in price, but very inferior in binding and typography and without the paragraph headings that are so useful. Our edition is complete with all these improvements, is beautifully printed and exquisitely bound in cloth, and the retail price is only $1.00, with discounts to teachers and reading-circles.

SEND ALL ORDERS TO
E. L. KELLOGG & CO., NEW YORK & CHICAGO.

Analytical Questions Series.

No. 1. GEOGRAPHY. 126 pp.
No. 2. HISTORY OF THE UNITED STATES. 108 pp.
No. 3. GRAMMAR. 104 pp.

Price 50c. each; to teachers, 40c; by mail, 5c. extra. The three for $1.20, postpaid. *Each complete with answers.*

This new series of question-books is prepared for teachers by a teacher of high standing and wide experience. Every possible advantage in arrangement of other books was adopted in these, and several very important new ones added. The most important is the

GRADING OF QUESTIONS

into three grades, thus enabling the teacher to advance in her knowledge by easy steps.

THE ANALYTICAL FEATURE

is also prominent—the questions being divided into paragraphs of ten each, under its appropriate heading.

TYPOGRAPHY AND BINDING.

Type is clear and large, and printing and paper the very best, while the binding is in our usual tasteful and durable style, in cloth.

The books are well adapted for use in schools where a compact general review of the whole subject is desired. The answers have been written out in full and complete statements, and have been separated from the body of the questions with a view of enforcing and facilitating the most profitable study of the subject. The author has asked every conceivable question that would be likely to come up in the most rigid examination. There are other question-books published, but even the largest is not so complete on a single branch as these.

Bear in mind that these question-books are absolutely without a rival

**FOR PREPARING FOR EXAMINATION,
FOR REVIEWING PUPILS IN SCHOOL,
FOR USE AS REFERENCE BOOKS.**

The slightest examination of this series will decide you in its favor over any other similar books.

Love's Industrial Education.

Industrial Education; a guide to Manual Training. By SAMUEL G. LOVE, principal of the Jamestown, (N. Y.) public schools. Cloth, 12mo, 330 pp. with 40 full-page plates containing nearly 400 figures. Price, $1.50; to *teachers*, $1.20; by mail, 12 cents extra.

1. *Industrial Education not understood.* Probably the only man who has wrought out the problem in a practical way is Samuel G. Love, the superintendent of the Jamestown (N. Y.) schools. Mr. Love has now about 2,400 children in the primary, advanced, and high schools under his charge; he is assisted by fifty teachers, so that an admirable opportunity was offered. In 1874 (about fourteen years ago) Mr. Love began his experiment; gradually he introduced one occupation, and then another, until at last nearly all the pupils are following some form of educating work.

2. *Why it is demanded.* The reasons for introducing it are clearly stated by Mr. Love. It was done because the education of the books left the pupils unfitted to meet the practical problems the world asks them to solve. The world does not have a field ready for the student in book-lore. The statements of Mr. Love should be carefully read.

3. *It is an educational book.* Any one can give some formal work to girls and boys. What has been needed has been some one who could find out what is suited to the little child who is in the "First Reader," to the one who is in the "Second Reader," and so on. It must be remembered the effort is not to make carpenters, and type-setters, and dressmakers of boys and girls, but to *educate them by these occupations better than without them.*

SEND ALL ORDERS TO
E. L. KELLOGG & CO., NEW YORK & CHICAGO.

Augsburg's Easy Things to Draw.

By D. R. AUGSBURG, Supt. Drawing at Salt Lake City, Utah. Quarto, durable and elegant cardboard cover, 80 pp., with 31 pages of plates, containing over 200 different figures. Price, 30 cents; *to teachers,* 24 cents; by mail, 4 cents extra.

This book is not designed to present a system of drawing. It is a collection of drawings made in the simplest possible way, and so constructed that any one may reproduce them. Its design is to furnish a hand-book containing drawings as would be needed for the school-room for object lessons, drawing lessons, busy work. This collection may be used in connection with any system of drawing, as it contains examples suitable for practice. It may also be used alone, as a means of learning the art of drawing. As will be seen from the above the idea of this book is new and novel. Those who have seen it are delighted with it as it so exactly fills a want. An index enables the teacher to refer instantly to a simple drawing of a cat, dog, lion, coffee-berry, etc. Our list of Blackboard Stencils is in the same line.

Augsburg's Easy Drawings for the Geo-

GRAPHY CLASS. By D. R. AUGSBURG, B. P., author of "Easy Things to Draw." Contains 40 large plates, each containing from 4 to 60 separate drawings. 96 pp., quarto cardboard cover. Price 50 cents; *to teachers,* 40 cents; by mail 5 cents extra.

In this volume is the same excellent work that was noted in Mr. Augsburg's "Easy Things to Draw." He does not here seek to present a system of drawing, but to give a collection of drawings made in the simplest possible way, and so constructed that any one may reproduce them. Leading educators believe that drawing has not occupied the position in the school course heretofore that it ought to have occupied: that it is the most effectual means of presenting facts, especially in the sciences. The author has used it in this book to illustrate geography, giving drawings of plants, animals, and natural features, and calling attention to steps in drawing. The idea is a novel one, and it is believed that the practical manner in which the subject is treated will make the book a popular one in the school-room. Each plate is placed opposite a lesson that may be used in connection. An index brings the plates instantly to the eye.

Song Treasures.

THE PRICE HAS BEEN GREATLY REDUCED.

Compiled by AMOS M. KELLOGG, editor of the SCHOOL JOURNAL. Beautiful and durable postal-card manilla cover, printed in two colors, 64 pp. Price, 15 cents each; *to teachers,* 12 cents; by mail, 2 cents extra. 30th thousand. *Write for our special terms to schools for quantities. Special terms for use at Teachers' Institutes.*

This is a most valuable collection of music for all schools and institutes.

1. Most of the pieces have been selected by the teachers as favorites in the schools. They are the ones the pupils love to sing. It contains nearly 100 pieces.

2. All the pieces "have a ring to them;" they are easily learned, and will not be forgotten.

3. The themes and words are appropriate for young people. In these respects the work will be found to possess unusual merit. Nature, the Flowers, the Seasons, the Home, our Duties, our Creator, are entuned with beautiful music.

4. Great ideas may find an entrance into the mind through music. Aspirations for the good, the beautiful, and the true are presented here in a musical form.

5. Many of the words have been written especially for the book. One piece, "The Voice Within Us," p. 57, is worth the price of the book.

6. The titles here given show the teacher what we mean:

Ask the Children, Beauty Everywhere, Be in Time, Cheerfulness, Christmas Bells, Days of Summer Glory, The Dearest Spot, Evening Song, Gentle Words, Going to School, Hold up the Right Hand, I Love the Merry, Merry Sunshine, Kind Deeds, Over in the Meadows, Our Happy School, Scatter the Germs of the Beautiful, Time to Walk, The Jolly Workers, The Teacher's Life, Tribute to Whittier, etc., etc.

www.ingramcontent.com/pod-product-compliance
Lightning Source LLC
Chambersburg PA
CBHW030010240426
43672CB00007B/896